Where We Are

We live in a fast-paced world full of monstrous contrasts. Certainly, the contrasts were there during the time of the Pharaohs, according to our history books, but they were very different. Some of them were hungry and worked hard, while others were well-fed and relaxed. Today we experience radically different contrasts, when hungry people still have to suffer due to hunger and do manual labor to feed themselves and their families. Not only do the well-fed not work, but they use all sorts of technical marvels to ensure that they will never have to work again and, in addition, to avoid manual labor at all.

The enormous pace of technological progress was probably the main reason for these contrasts. The gap in living standards between people in developed and developing countries is astonishing. However, people themselves are not inherently very different, and the only reason for such a gap is probably the difference in the initial natural conditions. I proceed from the premise that people of all nations, nationalities and races are basically the same and when placed in specific conditions, they will change accordingly, regardless of their origin. I would also like to specify that within the scope of this research, the term "developed country" is primarily characterized by the level of automation and robotization of various kinds of processes occurring in such a country. In order to understand my train of thought, examples of such countries include Germany or South Korea. But I do not want to go into the specifics of any particular country in this book, so as not to get distracted by particulars. Obviously, each country has its own achievements and problems, and so as not to evoke unnecessary emotions in the readers or offend their patriotism, I will leave specific examples out of the equation and limit myself to considering the processes, technologies and their transformation.

We cannot fail to note one important issue for further discussion. The fact is that the overwhelming majority of people nowadays spend more and more time on their smartphones. I believe I can specify the main reason for this without providing excessive explanation or evidence.

So, apart from its obvious benefits, a smartphone today enables a modern person to escape from reality almost completely and move to the world of the Internet, virtual communications, etc. And people love it, they want it. The technologies that provide such close contact with what we shall hereafter call the virtual world of communication have been actively developed over the past 30 years. However, their real significance for further changes is only now becoming clear to us, in the context of the coronavirus pandemic, following

Introduction

And there's no other way
To make everyone happy,
But to put them
Into a fictional world.

What will our world look like in 20 to 30 years' time? The short answer is it will become virtual. In this little book we will figure out how and why, touching upon the main spheres of human life and tracing the changes that may happen to them.

I would like to warn the reader in advance that everything written here is solely my own review of current events and, of course, I cannot claim to be in possession of an absolute truth. Not being a true professional in any of those areas of change discussed in this book, I have been trying to look at each of them from the perspective of an ordinary person. I was inspired to write this book because I felt the need to consider my next steps, to choose my further area of activity and to answer the question of what our children's world would be like. My great desire to get accustomed to this new world became the main reason for writing. It was quite difficult to formulate thoughts in my head without giving them a   structure. Moving from one page to another and opening up the prospect of different technologies with regard to the main spheres of modern human existence, I tried to form a coherent picture of the life that would be inherited by our next generation. No doubt we will be within that picture and will be part of it, but our descendants will not be able to comprehend the changing dynamics of the past twenty years and those changes that will occur in another 20-30 years.

We managed to catch the emergence of ground-breaking technologies. Our ancestors witnessed the launch of the first steam engines, but they could see the whole process with the naked eye and understand every little detail guided only by their intuitive grasp of fundamental physical processes.

I would also like to note that while making assumptions and forecasts I have not tried to imagine the consequences of global warming, nuclear war and other similar phenomena, which have already been actively discussed for quite a long time. Thus, we will assume that in 20-30 years the impact of such factors will remain at current levels of 2020. Otherwise, our civilization can seriously suffer and set back its development by several decades or radically change, and then everything written below will lose its meaning, partially or even completely.

I am writing this book in April 2020, during the height of the so-called coronavirus pandemic and the resulting economic crisis, which most analysts predict to be the worst in the last hundred years. We will know soon enough if they are accurate in their predictions. However, in any case, significant changes in the world are supposed to happen. After all, technology has made great strides over the past decades, but people have hardly changed.

Before I start, I would like to point out that those changes I am going to describe are not strictly related to the latest technologies of the future. It is far more important to me to highlight the key processes that are already starting to change imperceptibly under the impact of these technologies, and to observe the possible consequences. Technological developments outlined in this book are based on the analysis of already existing trends in automation in various industries, along with the prominent scientific discoveries and research that may point to possible trends in the coming years.

The book is not intended for those who are familiar with current technologies and can clearly see what will happen to our world in decades. Much of what is written below may seem obvious to them, so they should not waste their time reading it. For everyone else, I hope it will be interesting to learn what awaits us in the near future.

This is my first book, and it probably contains numerous shortcomings. Perhaps some chapters are not as detailed as the others. But as they say, time is running out. In preparing the next edition, I will certainly make the necessary corrections and additions. In the meantime, let's read about the things that have already started to happen to us - because we can still become active participants in these events.

many days spent in quarantine. For our research it is entirely irrelevant how real this threat has been and continues to be, but under its influence we have gained a new understanding of the opportunities and consequences that come with the vast majority of people having easy access to the virtual world of communications.

Over the past 20-50 years we have never once found ourselves massively (city-wide, country-wide) in a situation where we actually couldn't leave the house for more than a month in a row (apart from short trips to the shop or pharmacy). And technologies have developed rapidly all this time. Now they will make us realize that we are ready for fundamental changes in all spheres of our lives without exception.

What's Going to Die First

Clearly, we cannot predict the exact date when a certain technology will be introduced worldwide, causing some processes to die out. But to consider our assumptions as coming true, we only need to see it happen within at least one city in a single country. The rest will gradually catch up.

It would be easy to start with those areas where disappearance or total change are already obvious and clearly long overdue. We will give every consideration to them below, but here we will share some generalities. While access to the virtual world of communications already enables most functions to be performed remotely, due to a habit formed over previous decades (or perhaps centuries), their implementation would have involved, until recently, a physical contact. Nevertheless, certain people's fear of being presumably infected and the widespread imposition of quarantine for quite a while became a strong catalyst for the transition to new approaches in implementing these functions.

It has long been clear to everyone that it is completely unnecessary to buy a carton of milk in person in the supermarket since it makes no difference who exactly buys top-quality milk: you personally or a delivery man. And now, being in quarantine, we can be definitively sure that most foodstuffs are of pretty sustainable quality and can be ordered from the same supermarket online for home delivery. Even farm products from distant markets have long been ordered online or by phone in many countries. But now when the quarantine is over, the process of buying food and goods from offline shops will begin to fade away more obviously. Children of new generations will no longer understand why their parents go shopping. After all, there are more important and interesting things to do. And then, at some point, there will be no shops left in the street.

Another example is meetings of all kinds. It has been awhile (more than 10 years ago) since all the necessary online meeting and web conferencing tools, as well as automatic record-keeping systems, etc. were introduced. However, people used to shake hands and communicate in person, and this has been going on too long. However, now, after working remotely for weeks or months in a row, not just across organizations but across industries, people suddenly realize that it is much easier, more efficient, logical and safer to hold such meetings remotely. And this will become the new standard once and for all. In 20 years time, our children will not even consider the option of meeting in person, preferring to use the virtual world of communications. Moreover, in 20 years, we will almost always communicate with our children only remotely, although much more frequently than now.

Also, physical luxury goods segment will die out almost completely. Once people stop communicating in person, any physical attributes of superiority will no longer matter, as any image can be created in the virtual world at a much lower cost. In 20 years, expensive watches, suits, cars and jewellery will seem like irrelevant props.

Will we be using paper documents in 20-30 years' time? Will we be using paper at all? It seems very unlikely to me. It might be used in small quantities for gift books and some other products that will be used mainly by adults and the elderly. As for paper money, I am sure that even underdeveloped countries will realize that changing the law and introducing new technologies is cheaper than continuing to cut down forests and subsequently wasting time and space on storing endless boxes of documents that no one will ever read again. Paper money will also disappear.

In 20-30 years, all taxi driver jobs will disappear. The next generation may not understand why they need a steering wheel. The steering wheel will disappear, and so will the pedals.

In general, all mediation and almost all professional activities will be completely gone in 20-30 years. Total automation and cross-sectoral integration will do the trick.

What Will Change

Let us highlight and list the most important areas of our lives that will change over the next 20-30 years, and then, in the following chapters, focus on each of them in more detail.

Information technology. The rapid development of information technology will continue to be a major factor influencing large-scale changes in all other industries. The IT sector itself will search for new opportunities to automate and accelerate its key processes, while actively taking over other industries.

Telecommunications. Telecommunications will be among the first industries to actually merge into the IT sector. Telecommunications companies will redesign their processes to resemble IT companies as much as possible, providing the necessary speed to bring new products and services onto rapidly changing markets.

Energy. Energy will become much more renewable, and there will be a variety of ways to generate energy.

Transport. The transport sector will change significantly. It is no longer a secret that self-driving cars will soon fill the streets, and that will enormously help accelerate progress in general.

Trade and logistics. Trade will turn into logistics and logistics will turn into trade. The consumption pattern will change dramatically, which will result in restructuring the product range.

Real estate and construction. New technologies will redefine attitudes to real estate in general and enable people to strive for more than just purchasing a small flat on credit. The time will come to intensify the processes of de-urbanization. Construction will become fast.

Household appliances and electronics. In the world of home appliances and electronics, we will meet new generations of familiar devices. New ones will also appear. Home life will become even more comfortable and interesting.

Agriculture and the food industry. The agricultural and food industries will try to feed us with the cheapest products full of various chemical additives. Alternatively, we will have autonomous small farms producing better quality and more healthy food.

Education. Education will become completely remote, providing everyone with huge opportunities for self-development. However, an approach to requiring secondary, let alone higher, education will be changed, so that fewer people will study hard.

Healthcare. Medicine will significantly redesign its processes and automate them, shifting the focus to collecting and processing tests and replacing live doctors with automated systems wherever possible.

Sports. Sports will use the latest technologies to greatly equalize opportunities for totally healthy people and people with disabilities.

Tourism. New trends in tourism will emerge as virtual reality technology develops. Also, new tourism destinations will be actively developing.

Hospitality. Hotel and restaurant businesses will be completely transformed. The demand for hotels and restaurants will gradually decrease and all surviving players in the market will adjust to the new trends and habits.

Entertainment. Entertainment will move almost entirely into the virtual world of communications and give us new emotions.

Services. The services sector will be automated and robotized wherever possible. Many services will either disappear or be completely transformed.

Advertising and marketing. Advertising and marketing will follow us directly into the virtual world and haunt us there, imperceptibly turning from annoying neighbors into helpful assistants.

The media. The media will obviously make a complete transition to the virtual world of communications. They will change the presentation of materials, and censorship will work differently.

Financial institutions. Financial institutions, such as the telecommunications sector, will begin to turn into IT companies, adjusting to the large-scale digitalization of the economy.

Business as a whole. Once the current crisis is over, many companies will certainly try to return to their former lives and restart their processes under the same configuration. However, new young businesses will begin to actively displace them, adapting faster to the new vectors of the next generation's consumption patterns.

Military. Armies will start to downsize. Humans will be replaced with automated machines. Military conflicts will change in nature, so as to save as many human lives as possible.

Security. Security will become absolute – there is nothing more to add.

Space exploration. Space exploration will proceed very smoothly waiting for key knowledge and technology that would transport us (or at least our consciousness) to other galaxies.

The role of the government. All countries will be led by people born in the world of mobile phones. After that, the state's role in the development of technology will become decisive.

Separately, I would like to comment on human chipping, which is an increasingly debated and, for some people, crucial issue. We have been chipping dogs for a while now and it seems logical that we should start chipping people as well. Certainly, there are pros (in terms of life automation, for example) and cons (primarily, perhaps, a restriction of freedom). In my opinion, it makes no sense at all to argue about whether we should chip people or not. No one under the age of 40 (with very few exceptions) is now able to get away from a mobile phone, which contains all the vital information. Phones are certified by the intelligence services and can easily be monitored by them. From this perspective, chips will be no different. I think in the next 20-30 years, people will still have a choice whether to be chipped or not, but then it will happen everywhere, as it did with mobile phones. In fact, in 20-30 years, I believe the mobile phone will disappear from our lives, and it will be replaced by chips and new virtual reality technologies, which will let us communicate in a slightly different way than we have been used to since the first phone was invented.

In conclusion, we should, of course, ask ourselves: in 30 years' time, will there be specific answers to the main questions that have confronted humanity since its inception? Or will new questions arise and old ones gradually be forgotten?

Information Technology

New era of information technology started about 25 years ago with first powerful personal computers, user-friendly operating systems and first mobile phone prototypes. At that time, few people expected the current pace of IT development. If you visited the office of almost any company (whether a factory or a bank) in any country, you would be unlikely to find many IT staff.

You could at best see a system administrator fixing computers, mice and printers, and in particularly difficult cases - even an entire IT team of three to five people. However, all these people performed routine operations, and if there were a lot of them, it was only due to the large number of computers and office equipment. By the way, at that time IT specialists were considered to be second-class people, part of the back office, people who could increase company costs.

Today, approximately 5 years later, all information processing organizations have started to transform, gradually acquiring more and more office equipment and computers, and introducing various software products to automate their business processes. As a result, IT teams turned first into IT groups and then into IT divisions, and then into IT departments. Automation costs have increased several times over, and information technologies have become a key component of the corporation's global business strategy. Of course, in some industries this process was faster than in others. Many companies continued doing things the old-fashioned way, reluctantly using IT and only in processes where they were de facto already a standard.

Today we can hardly imagine an organization of any scale or an industry that would not use information technologies in its activities. And not just use it. The situation has changed so much that today all key business processes contain mandatory automation elements. In large organizations, IT managers quickly became stronger in the management structure. When launching new business lines or companies, business owners first seek advice from IT directors. That can be easily explained. All these years, IT specialists have worked at the interface of business and technology, which made it possible to understand the specifics of the business, formalize all algorithm-based business processes, learn to understand major business objectives and key performance indicators.

At the same time, representatives of business units, by inertia, considered IT to be something subordinate and thus hesitated to set automation goals, not interested in studying the architecture, information processing, etc., which in the end played a nasty trick on them.

Even now, IT people are more likely to create new businesses, organization leaders try to listen to IT first, and this is absolutely justified. Over the past 20 years, major industries and business areas have changed much less than IT, no matter how strange it sounds. Loans were issued 20 years ago, but now they are issued in a completely different way, and only due to information technology. New ways of interacting with customers, information processing, etc. have appeared. IT-professionals, doing their utmost to ensure business development, immersed themselves into business specific features and began to offer new business processes based on their own initiatives, understanding, and the best approach to business automation in each case.

We have paid so much attention to IT transformation in order to draw your attention to the amazing speed at which it happens. These changes are only the beginning of complete reconfiguration of various enterprises in all industries within the next 20-30 years. That's the reason why this chapter is first on the list.

It would be no exaggeration to state that in 20 years any company will be led by a person with deep knowledge of IT. I would venture to guess that almost all new leaders will come from IT departments of their own or similar companies. In the coming years, most corporate competitive advantages will be based on proper automation of business processes. No new business ideas will be implemented without initial detailed design of their automation.

Let's move on. How will IT companies change in the next 20-30 years and what will they become?

If you look at our world today from the year 2050, IT products and technologies that we use these days may seem not so advanced to us. We read and watch science fiction and expect the IT world to offer fundamentally new solutions. High expectations are related to the so-called artificial intelligence, which promises considerable transition to a new world in the near future.

In my view, the main challenge of modern information technology is the transition to a new technological era of automating software generation based on verbal or visual initial objective setting. In other words, we need that technology to automatically collect data, in whatever form, in a "man-machine" mode and immediately acquire the software prototype of the required product for further discussion and improvement. It should take just a few seconds, including all kinds of testing, to generate program code. In the near future, large IT companies will create solutions that will facilitate the business process automation.

On the other hand, IT companies will probably aim at transforming themselves into technology companies to produce not only impersonal software products, but also specific devices (in electronics, robotics, medical appliances, military equipment and other industries). This will enable to affect the outcome and modify the devices during the manufacturing process. Moreover, all device prototypes (as well as mass production in the future) will be manufactured using 3D printing and will not require such high production costs as before. It seems likely that after the market launch of more powerful and functional 3D printers, IT companies will simply take over the entire market of next generation smart devices, since electronic "guts" of any device will be identical to the "guts" of a modern laptop, unless they may become even smaller and more efficient.

What changes will happen to the software products produced by IT companies? It used to take too much time - weeks, months or even years - to create and finalize any typical product. I believe this situation will be rectified soon. The business community is ready to criticize IT specialists for not acting promptly with regard to effective automation. We've come to the point where we can instantly change any business process and immediately start making additional profits. Of course, this will require people with the right skills, able to speak machine language who can rapidly make changes to the process and at the same time receive the required program code.

We can assume that very soon all typical products will be equipped with a special interface enabling you to quickly explain to the system what to configure according to the specifics of your business. The improvements will be automatic due to a built-in "engine". Such products will conquer the market as soon as they appear and become a new standard of software development. We need a fresh look at how to teach a machine to understand a person and adjust its software product. Perhaps, the first developments in automatic system configuration will be associated with piloting this system by the most advanced users, who, by their actions, will make the system understand what they really want from it, that is, to teach it. As for artificial intelligence, it will help adapt the system to the needs of the organization.

Looking closely at the IT market, you may see that some industries have long been automated, including specialized IT companies. Examples of such industries are healthcare, logistics and others. Of course, even these industries are far from being mature in terms of a sufficient number of specialized players on the IT market. In the coming years, IT industry will move towards further intersectoral stratification. Modern IT companies are typically seeking to collaborate simultaneously with companies from various sectors, which is primarily due to very small budgets for automation. There is a growing perception of what real automation should be, and that explains why business (in broad term) is trying to reduce costs for automation.

Organizations should make a strategic decision to move closer to remote operation of each division, which will expand the boundaries in the literal and figurative sense. As soon as the leaders of all types of organizations realize that from now on their activities should become as decentralized as possible, business processes will be completely reconstructed and will approach maximum automation and robotization. Business decentralization essentially means being able to produce anywhere (lower costs), sell to anyone (regional expansion), and work from anywhere (mobility). And that creates huge business growth potential for those companies which will be the first to implement the restructuring and automation. For these reasons, the IT market expects further growth and, as a consequence, further stratification.

The restructuring and automation of business processes in several branches will demand the special equipment ensuring robotization of certain functions. Since automation and robotization should be inextricably linked in the future, as we have already mentioned, IT companies can and should begin to transform into technology companies, ready to offer not only supply and configuration of various equipment and software, but also to create the necessary equipment from scratch.

One more trend in IT is access to non-specific markets through acquisitions of companies representing different branches of business. Those companies, which have to be confronted with reduced profitability and cannot make quick changes, will be targeted by IT specialists who have a thorough understanding of how to quickly automate and robotize key business processes. The main interest will be to gain access to specialists familiar with sectoral specificities and to inherit some elements of the manufacturing base. Once the acquired business with the new technological processes is resumed, the new owners will increase their companies through further acquisition and deeper automation. In the following chapters of the book, we will consider, through concrete examples, how the above trends are going to work.

Telecommunications

Mobile telecommunication services and the first mobile phones appeared on the market about 25-30 years ago. Back then, mobile communications were characterized by mobile coverage. Cell phone towers were quite actively built, and soon people in large cities stopped paying attention to this parameter - they believed that in general, communication was everywhere. Mobile Internet service along with first mobile applications for download proved to be a real breakthrough. Immediately afterwards, it became clear that the line between the personal computer and the telephone would soon blur, and this proved to be the case.

In all years since then, nothing special has happened. Although mobile network operators continued to move inland, expanding their coverage, creating new standards for better quality service, there was no global change for an average user. With the introduction of various messengers, we can save money using the Internet to call, instead of a phone line. However, mobile network operators, without providing ordinary users with new principal functions, actively collected various information and cooperated with the government and businesses by creating new services for them. Today mobile network operators can completely control our lives, from determining our location to receiving all information transmitted through a mobile phone.

We need to understand that initially telecommunications companies were mainly infrastructure companies, providing some kind of pipes for pumping information through them. Over time these pipes became wider and wider. Applied technology has not been the main priority all along just because the width of the pipes and the overall coverage of the territory have not reached its maximum. This means that various telecommunication companies have continued to struggle vigorously for their main market.

What awaits us in another 20-30 years? Firstly, very soon all remote areas will be provided with full satellite coverage, which will remove the problem of white spots on the network coverage maps. Secondly, next communication standards are coming up, which will fundamentally increase the cross-section of pipes. Wireless Internet connection speed will become so high that the need for wired Internet will disappear once and for all with next generation coverage. And that's when we will see the rapid transformation of many telecommunications companies into full-fledged IT companies. Having full control over the entire infrastructure, which enables users to instantly exchange any amount of information over the air, they can get rid of all other market participants in one fell swoop by blocking their services if necessary. Not all telecommunications companies will probably go this route, but they will definitely have the chance.

Satellite Internet and the incredible growth of wireless Internet speed lead us provide much food for thought. It seems to me that cloud storage has not yet utilized to its full potential. These days, the term is mostly used figuratively, because the data is actually stored (and processed) on remote physical servers, and not on the end user. But real clouds cannot be on the ground. And therefore, at some point, when new telecommunication standards will provide an almost unlimited speed of wireless data transfer, we will get the cloud Internet - the pipe size will enable huge amount of transmission and storage of data, as much as we need, because these pipes will not only be wide, but also layered in some sense. When that happens, we will have incredible opportunities to use information. And then we will achieve true cloud computing technologies, for example, by creating super-powerful servers on satellites that will not heat up so much in space. In any case, these servers could also be placed on cell towers. And this also offers great opportunities for the telecommunications sector. It won't take more than 30 years to implement the technologies themselves, given how quickly we reached our current speeds, starting from zero twenty years ago.

However, speaking further on behalf of the satellite Internet, some countries are already actively launching their space vehicles into space, while others have been inactive. This may lead to the fact that pioneers will have time to build a fundamentally new communication infrastructure that may cover the entire planet (both on land and on water) and force those countries which are lagging behind to use it. In this case, the entire land communication infrastructure will become secondary, performing as a kind of last mile. And this only confirms the importance for local telecommunications companies to search for new business niches, if they are incapable of fighting for space.

Why do telecom companies and why do we need such a dramatic increase in communication speed? It goes without saying, we are moving towards the era of the Internet of Things, when every electronic device will be connected to the Internet by default, but the current channel size will be enough for that. However, it's not so much about the emergence of a large number of new potential users, but a qualitative change in the approach to communication between people and information exchange.

We stand at a very important point when new technology is about to reveal the concept of true virtual reality. If 20 years ago it felt like a miracle to have access to the Internet anywhere, now we should brace for instant transmission. High-speed communication will enable moving full three-dimensional copies of any part of space to a given place, simulating reality and allowing a large number of people and other objects to remain in this reality at the same time. Naturally, we will need special devices to get into this virtual reality. We'll talk about it in the "Household appliances and electronics" chapter.

And again, telecommunications companies will be able to take control of the entire market if they want to, creating various virtual worlds and developing special software to work with them. These worlds can be completely fictional, or dynamically copied from the reality by using next-generation video equipment, or a combination of both. We shall talk further of how these technologies will be used in practice. It is evident that telephone calls can turn into a full-fledged conversation between people anywhere with access to the corresponding virtual world. In fact, these virtual worlds can become another kind of infrastructure in terms of telecommunications companies - one of the ways of intensive business development.

Will SIM cards be phased out in 20-30 years? Certainly, corresponding technologies have been ready for a long time and are already being used in some countries. However, for the vast majority of states, a SIM card is, in fact, a mechanism for monitoring citizens, uniquely identifying them and their location. And few people in the near future will be ready to abandon this tool, which has existed for several decades. Moreover, instead, soon more and more functions (including a simulated passport) will be added to the SIM card. There is some suggestion that this is already occurring little by little. We will probably be provided with new modifications of such SIM cards until they are replaced with computer chips and installed in our brain.

Chip implantation will reveal endless opportunities. Access to the virtual world will be permanent and will not require additional devices. Our body will supply power to the chip. To talk to another person, you no longer need to open your mouth, since chips will immediately read articulatory vibrations, which will also be used to form the appropriate information flow to your interlocutor, keeping your tone and all intonation. However, I guess, this may go beyond the time frame of 20-30 years, although in the foreseeable future, and we will talk about it in detail.

Energy

All aspects of living and all spheres of the economy are completely dependent on energy, because everything needs energy. But in the coming decades, the energy sector itself will depend on the development of information technology and telecommunications.

It happens that, as I write these words (April 21, 2020), the price of oil in the financial markets has hit another historic low. Does this mean that all attempts to introduce alternative energy will be blocked by low oil prices? Why look for something new when there is a familiar way that is cheap again? But no, wind and solar generators will remain in place. For too long we have relied entirely on black gold, since very entrepreneurial people from the previous century were able to take complete control over the energy sector and dictate their terms to the markets. What is the reason for the sharp rises and falls in energy prices? As an outside observer, I would venture to suggest that the core of the problem is that the world's energy needs are volatile and depend on different factors. Previously, many believed that the energy demand could only increase over time, but no - it could also decrease. It can decrease not only due to the epidemics, pandemics or ordinary economic crises, but also due to the invention of new technologies that consume significantly less energy compared to current parameters. And then, storing the already saved energy will become an unbearable burden, and energy prices will fall, even when everyone is healthy and regularly goes to work.

On the other hand, current oil production methods do not always allow quick and easy control of energy production. In other words, it is not easy to quickly increase or decrease the amount of oil produced. In addition, for decades the economies of some oil-rich countries have been, and continue to be, tied to supplying oil to world markets. We find it difficult to understand the main factor preventing a more flexible approach to oil production. But the situation with natural gas seems to be similar.

I would dare to suggest that we are on the verge of some sort of energy revolution. In terms of technology, we have long been ready to produce as much energy as the world needs, and modern technologies already provide significant savings and are just beginning to gain speed of development and distribution. This means that global energy demand will continue to decline even as the world's population grows.

Developed countries are ready to completely change their electricity consumption patterns. In general, developing countries lack only one step - demonopolization.

Modern wind or solar energy production technologies can now fully cover the world's energy demand. But they are only in their infancy and therefore have great potential, especially when considered together with technologies for reducing consumption, redistribution and shared storage and consumption of energy.

How will the energy consumption model look like in 20-30 years' time? Provision of independent electricity generation will be taken into account prior to design of residential houses and any other kinds of structures. Some areas will use wind turbines, while others will use solar panels. In some locations, there will be other technologies which have also been around for quite a long time, but are not yet widely used. Already constructed buildings will be converted, and various power plants will be installed on unused land. The main requirement will be to use only Green energy. Most households will generate their own energy through renewable sources. The surplus of electric energy will be stored and transferred to general distribution centers, where energy can be stored and redistributed to facilities suffering from power shortages. Industrial and other enterprises will require more energy than they can produce on their own in their territories and facilities. Thus, they will probably use excess energy. The process of redistribution itself can be of economic nature depending on the role that the government will take on.

In this chapter, I am not going to dwell on fundamentally new green technologies, because I am not an expert in this field. What I mean is, one must realize that in the coming years more and more new solutions are going to enter the market, enabling to produce, accumulate, store and redistribute environmentally friendly energy that are meant for various consumers. Currently, in many countries it is hampered by the existing rules and the lack of understanding that the time has come.

Obviously, energy storage and distribution will be fully automated, and energy will be handled just like any other product. Information technology will assist here. New automated systems will ensure that energy derived from all producers is taken into account and will be able to control all its consumers.

All the above is pretty clear; we see more and more wind turbines and solar panels. We just have to wait until their share will reach a critical level and traditional energy sources begin to gradually disappear from the market. However, electric cables are another element of the energy production and consumption model, which requires significant changes.

In developed countries, we do not usually notice any electrical cables in the streets, because they are usually underground. In developing countries, the cost of laying underground cables is often reduced, and a traveler may see an incredible web intertwined between poles and various structures. But it won't always be that way. Solutions that can "cut" cables for power transmission are no longer secret. In 20-30 years, we will see the first areas or cities completely cleared of electrical cables.

First, wireless electricity will be used inside buildings, as distances will not be a significant delaying factor in the process of transition to new technologies. All devices will be connected to a wireless network that provides both the Internet and electricity.

With new opportunities to extend the distance for wireless power transmission, electrical cables will begin to disappear from the streets. This means that absolutely all equipment will always be connected to new generation power grids. At the same time, electrical grids will contain information on all the areas with reduced or increased energy supply and redirect electricity to where the demand is higher. This may sound quite trite, but we can't overestimate the value of switching to wireless electricity. All electrical appliances will not only get energy from the grids, but also return it back and exchange it with each other.

Thus, in 20-30 years, energy and information will become a single whole. When we receive information, we will receive energy, and vice versa. And then telecommunications companies will be able to find new markets using existing infrastructure to upgrade their transmission capabilities along with information.

Transport

Over the next two decades, rapid changes in the transportation industry will dramatically alter other industries without exception.

Undoubtedly, the development of transport will be associated with two trends: robotization and car sharing. As a result, we'll get a robotic car sharing model. This will take no more than 20 years, at least in more developed countries. Let's see how this model will look in practice.

Imagine, you need to travel somewhere outside your area. To do this, you just need to call an unmanned vehicle via a mobile application. The car will drive up to your house and either will run along the nearest streets until you leave the house, or it will park in an empty parking space. Since the number of private cars will gradually decrease, parking almost everywhere will not be a major problem. As soon as you signal that you are approaching, the car will drive up to you as close as possible and unlock or open the door as you approach (depending on the model). When you get in the car, it will lock (or close) the door and drive, providing you with detailed information on your trip along the way - total travel time and cost. Of course, during the trip, you will be offered music and videos, water, the price of which will be added to the trip, and maybe something else. As you approach your destination, you will be asked if you need a car to wait for you there, in case you plan to return soon. In this case, the car will do what it did when it arrived on your call: it will either park in the nearest parking lot, or it will run around waiting for your signal. Certainly, the price of the trip will be charged from your account automatically, as it is already happening now. In addition to standard trips, cars will deliver goods or packages and perform other assignments.

Will the transport use gasoline or diesel fuel in 20-30 years? I don't think so. Developed countries will fully switch to electricity during this period. In developing countries, everything will depend on the government's position and financial resources of business. With enough charging stations throughout the country, the transition to electric cars will be very easy. In general, it is quite difficult to imagine a car without a driver using gasoline or diesel fuel.

What will the public transport be like in 20-30 years and will it exist at all? All our planes and trains will be completely unmanned (in many countries trains have long been moving without drivers). The number of buses, as well as other land transport, will decrease significantly due to no need for as many trips as today.

Now we can see massive traffic between cities or different districts in one city only because of the heterogeneous infrastructure and, in many cases, inability to work remotely.

Offices gravitate towards the center, so it is equally inconvenient for all employees to get to work. In 20-30 years, offices will be built on a totally different basis: they will be uniformly situated across residential areas and will not be owned or leased by any organization. On the contrary, they will be built and equipped according to a single standard and will allow any organization to rent the required number of workplaces, depending on the number of employees living in the area.

I assume that soon, more than half of office workers will work from home. Therefore, land public transport will not be in such demand. As for the underground public transport, it will become unprofitable and will be abolished someday. It will probably be meaningless to automate and robotize it because of small passenger traffic, considering the investments required for a complete reconstruction. Perhaps, if we start robotizing underground transport right now, there will still be time to recoup the investment. Nevertheless, these days we see some countries, on the contrary, actively developing underground transport, launching new lines and stations. But I can't imagine that, with all the trends in automation and such rapid advances in technology, passenger traffic will increase in the next few decades. Hopefully, transport companies don't wait too long to find out about reduced passenger traffic.

In some developed countries, we already see the airports of the future. There will be no human employees, just robots. Airports and flights should be fully automated to eliminate the human factor. You will arrive at an almost empty airport, within 5-10 minutes automatically perform all the necessary procedures and start boarding. Departure delays will be exclusively due to weather conditions, but you will also be warned about them automatically and in advance, and given no serious traffic jams and fully automated transport, you can wait at home to leave in time for the flight. I believe that eventually, airlines will take responsibility for door-to-door delivery of passengers, that is, airline vehicles will pick them up from a given location and then, upon arrival, deliver them to their final destination (office, hotel, apartment, etc.).

Will long-distance trains be in demand just like now? Today trains mainly carry people, coming to another city for a one-day trip or working during the week. In my opinion, in 20-30 years, this will not be necessary. To say the least, it seems weird that with at the current level of remote service development, companies spend their own funds on weekly transfers of employees from city to city, if not more often. In addition, these organizations pay for their temporary accommodation. Such expenditure seems completely unjustified and unreasonable burden on the transport system. In addition, with the advent of self-driving cars, all trips within 1500 km by car will become much more comfortable. Nowadays, many people prefer trains to cars mainly because they do not want to waste their time and energy on driving. And if you are traveling to a destination more than 1500 km away, flying might be a smarter decision.

Developing countries are facing challenging situations. Long-distance trains do not yet have the level of comfort that we see in the developed countries, and there are no good highways yet. But they will have to make a choice: to prepare the country for the era of self-driving cars and upgrade roads, or to invest in train refurbishment and high-speed rail construction. In any case, as I said before, we should expect a reduction in passenger traffic.

Will there be fundamentally new types of vehicles in the next 20-30 years, such as flying taxis or something like that? Once the traffic jams disappear, the need for such vehicles will no longer exist at all. It will be much easier to designate special vehicles that can move at a higher speed, or equip all vehicles so that they can use dedicated traffic lanes and have priority at intersections under certain circumstances, such as passenger's health hazard or other urgent need.

Traffic lights will completely disappear in a year or two after total ban on non-robotic vehicles on the roads. Including pedestrians.

Cars will perfectly "see" both pedestrian crossings and people.

In 20-30 years car trips from one city to another will be less frequent, but extra comfortable. People will enjoy perfect smart roads, an average speed of 200 km per hour, comfortable seats, full automation, excellent visibility and safety. Most vehicles on motorways will be unmanned trucks transporting goods. Many will travel at night to get a good night's sleep. Perhaps there will be some new class of vehicles equipped with sleeping chairs and used for interurban services.

As noted above, I'm not sure that in 20-30 years people will come up with some fundamentally different form of transport that can deliver passengers over long distances much faster than now, or that they will start using it massively. On the other hand, the comfort of such trips will enable people to see no difference between their usual day and the day they travel. Of course, all planes will have high-speed Internet, allowing travelers to stay in the virtual world of communications. For a person living at that time, being left without the Internet even for 5 minutes would be a real disaster. I am pretty sure, in 20 years' time, there will be no shuttle buses used to transport people to, from or within airports. The amount of time (up to two hours) saved at each of the stages will be especially evident for short distance flights.

It might be wrong to ignore individual short-distance vehicles. Their current fleet consists of bicycles, scooters and their electric successors (including gyro scooters, mono wheels, etc.). After switching to self-driving electric cars on all public highways, many people will miss the feeling of speed and perhaps some freedom. We will not now predict what will happen to modern motorcycles after the transition to unmanned cars, whether the automobile industry will preserve this type of transport or transform it into something else. However, an electric motorcycle is a fairly relevant and existing thing, but an unmanned electric motorcycle is something doubtful.

Therefore, individual short-range electric vehicles will show strong development. At the same time, the power and speed will increase after the safety systems of such vehicles reach the next level. Short distances, which used to be 20 or 50 km on a single charge, will grow to 200 or 500. Individual transport will continue to develop across the board; and people will be interested in getting maximum freedom of movement by land, air and water. To put it simply stated, very soon almost anyone will be able to climb Everest by such transport in a relatively short time. Accordingly, the infrastructure of the remote and virtually unvisited territories of our planet will begin to change. And we'll talk about it in more detail a bit later, in the "Tourism" chapter.

One of the other long-awaited changes in the transport sector in the next 20-30 years will be electrification and robotization in small aircraft. Flight speed and range shall increase, otherwise we will not see any practical benefits after the advent of self-driving cars. The next-generation light aircraft will have to fly at least 3000 km on a single charge at an average speed of 500 km per hour. In this case, we will experience a real breakthrough, given that the human factor will be reduced to a minimum, and given the slightest chance of an unfavorable outcome, the aircraft will look for better options to prevent the death of its passengers.

In conclusion, I would like to note that with regard to installation and updating of basic software, and use of additional applications, a car or any other vehicle of the future will be little different from a modern smartphone or computer.

Trade and Logistics

Trade and logistics in general are different sectors of the economy, but for a number of reasons, which will soon become clear, we will combine them into one chapter.

As I have already mentioned, conventional trading formats will soon begin to gradually disappear. People will tend to buy products and goods remotely. Supermarket and retail store owners will notice a decline in physical visits combined with an increase in online sales. Nevertheless, stores will continue to operate, gradually changing the forms of interaction with customers.

If you look closely at the work of modern supermarkets, you will see that they remind us more and more of a warehouse. Expensive renovated warehouses, located inside the premises with rents above market price. In such a warehouse, you can sometimes meet a uniformed employee who will definitely tell you where to find the desired product, but he can hardly report on its useful properties. I have often found that even in electronics supermarkets, the so-called salespeople (and in fact ordinary warehouse clerks) cannot explain or recommend anything, and therefore it is easier to find product descriptions on the Internet using data from the shelf sticker.

More than a decade ago, there were extensive discussions on the following issue. The authorities claimed that RFID tags in supermarkets would replace bar codes which could significantly speed up customer service at the cash registers and gradually eliminate the use of live cashiers. However, that technological innovation was underestimated despite attempts to implement it. It was also widely used in other areas where its application was more justified. Supermarkets have obtained more advanced barcode scanners and since then they have been actively adopting self-service checkouts. Next-generation trade equipment is so easy to use that any inexperienced person can make a purchase by himself.

Thus, we can say with confidence that supermarket owners will gradually, step by step, change our shopping routine, increase their profits and give up unnecessary business processes or shift the responsibility to us. All they have to do is to fire all the staff and move the storehouse to a place with lower rents. They can probably foresee how the world will change and are preparing for the least painful withdrawal of the retail formats we've grown accustomed to over the past decades from the market.

The first conclusion that we can draw from the above mentioned is that nowadays there is absolutely no need to visit supermarkets – it is high time to do the shopping online and to save time. Practice shows that when ordering online, the chance to get a low-quality or defective goods is not higher than accidentally take them from the shelf. One of the most important business processes of any self-respecting supermarket is quality control, which is aimed at tracking the feedback of online buyers and minimizing incidents with defective goods. The only reason we keep doing this is because we have developed this habit for decades. However, the next generation will no longer have this habit.

As people begin to realize that online shopping is more convenient, our consumption pattern as a whole will gradually change. All stores, without exception, will launch online sales in order not to lose their customers while streamlining business processes and reducing the number of sales consultants. Owners of all stores (including fashionable ones) will search and successfully find ways to provide customers with everything they need remotely. New mobile applications will help them to choose the right product, try-on it virtually and much more. Even the so-called "corner shops" will carry out a door-to-door delivery service to sustain sales and extend their existence. But they will also disappear one after another, giving way to next-generation vending machines.

However, the final understanding that there is no demand for physical stores will come only after the mass production of self-driving cars. By that time, a new generation of young consumers, unaware of the world without mobile phones and the Internet, will have grown up. These are precisely the people who will reinforce the tendency of moving away from any physical shopping.

At some point the construction of huge shopping malls will cease, and the existing objects will be converted. We shall return to this point later and consider what types of buildings they could be converted into. For now, let's examine how the buying process will change once unmanned vehicles hit the streets.

First, when retailers start using unmanned cars to carry out delivery to the customer's door, there will be a person who will monitor the car's progress and physically hand over the ordered goods to the buyer. This will enable to temporarily hire, for example, a large number of former taxi drivers who will start massively losing their jobs after the launch of self-driving vehicles. But we both know that the car itself can interact with the buyer directly, being a full-fledged robot with wheels. These cars are unable to deliver the purchases to the customer's door, which seems to be the only disadvantage. Due to this inconvenience, people will have jobs for several more years. However, we are now witnessing the launch of mass production of robotic assistants, which may eventually displace real delivery men. We'll discuss robotic assistants in more detail in the "Household appliances and electronics" chapter.

Certainly, a robot assistant will cost quite a lot. But if we talk about the possibility of recouping the costs within a year or two by replacing a living person and saving on wages, this will most likely not be an obstacle and we will definitely see one day robots bringing bags full of food in while self-driving cars are waiting for them.

The time has come for the second and the most important conclusion. We understand that launching unmanned vehicles will result in a sharp decrease in traffic congestion in cities for a number of reasons. This will allow supermarkets and other stores to improve the efficiency one last time before disappearing, moving their retail space to lower-rent areas and then eventually turning into warehouses.

Nowadays, supermarkets and shops have already become almost useless intermediaries in the process of transferring goods from the manufacturer to the buyer. The only advantage over conventional warehouses is their better location. However, as you may have guessed, this advantage will become insignificant with the wide-spread emergence of self-driving cars in the streets.

If there is an industry in our world that has been already automated and robotized as much as possible, then it is logistics. We do not even notice it simply because all the processes of delivery, storage and distribution are hidden from our view. Warehouses are located far away. So, delivery is mostly carried out at night time when the roads are less congested. However, if we were in a modern warehouse, we would hardly find people there. Logistics has long faced the need for full automation and robotization. Certainly, the society itself is to blame; people are consuming far too much.

If you ask me what logistics will be like in 20-30 years' time, I would say it will be all-encompassing. Logistics will absorb everything that is still ineffective in the market. First of all, it will absorb all trade.

We do not know if logistics companies are already getting ready to gradually occupy the retail niche, which completely depends on them (in terms of using their logistics capacities). But in any case, what remained to be done is not excessive. Like telecommunications companies, logistics companies already have the entire infrastructure to provide any customer with any product. As for now, the main obstacles are the current consumption pattern and traffic jams. In 20-30 years, both obstacles will be eliminated.

Obviously, large retailers are aware of the above mentioned. And, in any case, they are getting ready for these changes. Certainly, they have their own well-equipped logistics centers (hubs), and they are probably considering converting all shopping areas located near cities, as well as in residential areas, into robotic distribution centers.

Thus, small retail and all retailers who have not got their own logistics centers will be under attack.

Despite the fact that modern logistics centers have already become super-robotic objects, their finest hour is approaching. And by the time self-driving cars fill the streets, they will become fully autonomous, fully automated and robotic mini-cities without residents, where unmanned vehicles will stop for automatic loading or unloading. Companies that have already built or will build such centers will easily take over the retail market of any size.

Today, when we want to order a product in a supermarket with online delivery service, we visit the online store using our personal computer (laptop) or open a mobile application on a smartphone (tablet) and add products to the shopping cart. In 20-30 years, we will be able to make such purchases in virtual reality. It will be fun! We will discover goods and automatically receive all information on them. If necessary, there will be a virtual manager next to us, customized to our taste, who knows all our preferences and memorizes all our previous purchases. And most amazingly, we will be completely alone in this store. If we want to be among other people, of course, nothing is impossible in the virtual world. We can stay there as long as we want, and the manager will never get tired of talking to us and helping us shop.

Perhaps, I'm exaggerating a little bit, though, most people will just need to have the right stock of food at the right time and in the right place without too much trouble. In this case, it is enough to have a smart refrigerator that will order the missing goods in the supermarket and receive them directly from a robot directed by the logistics center on an unmanned vehicle. Only time will tell.

Real Estate and Construction

For more than one hundred years in a row, the cities had been growing up, and crowds of people were moving from the countryside to the big cities, leaving their homes and land. After all, life in big cities was exciting; people could earn more and live happier. Gradually, more developed countries leveled the infrastructure and some people left the cities. However, the level of urbanization is still extremely high and continues to grow in many countries.

What keeps us in the cities? All industries are moving to rural areas outside the big cities, but initially they were the ones that attracted people to the cities. Certainly, it is not that simple in real life, and apart from industries, there are a lot of things in the cities that are keeping people there now. However, during the quarantine, it turned out that many people were able to work effectively from home for weeks in a row, performing their standard functions almost without loss of quality. And this is due to existing tools of the virtual world of communications, but not to any special infrastructure.

Obviously, at least half of the city's residents are no longer tied to their workplaces. It will not be long before the owners and heads of organizations realize it and think seriously about reducing their office spaces.

What fundamental changes will take place in the real estate market in 20-30 years, apart from the obvious decline in demand for offices? The main trend will be "reurbanization." We will use this term to describe the process of outflow of some population groups from the cities and the simultaneous inflow of others.

It will start with hiring people living far away from the cities to work remotely in city offices, provided that such employees have the necessary infrastructure (Internet access and a personal computer or laptop) and good knowledge. We will enter a new era marked by great opportunities for everyone who wants to learn and develop. Remote learning, which we will discuss a bit later, will allow you to acquire almost any skills from the comfort of your home. And living conditions outside the big cities can potentially be much more comfortable. Detached houses in the country tend to have their own surrounding grounds and better access to wildlife than urban apartments, although house prices are lower. This means that working at home will be more comfortable for a nonurban resident, even if there are children in the family who, for some reason, should stay at home during the working day.

The overwhelming majority of the urban population now lives in fairly modest apartments, which often do not provide the necessary quantity of children's rooms, home offices for adult family members and guest bedrooms. City apartments are expensive, and sometimes it takes half your life to pay off your mortgage. The best half of our lives. However, it is believed that nowadays urban residents have more financial capacity and a higher level of education than rural residents.

Realizing that physical presence in an office does not have a fundamental impact on efficiency, companies will start cutting costs by getting rid of city offices. In other words, employees will have no choice whether to work remotely or not, as remote work will become a new standard. People will be forced to submit to the new reality, because with universal automation and robotization, it will not be easy to keep the job. Certainly, this will happen gradually and people will have time to adapt.

However, two important things will happen during the adaptation phase. Firstly, people in rural areas will be able to strengthen their position by getting remote jobs in the cities and occupying new promising niches. Secondly, urban residents, realizing that they are no longer tied to the cities, will start buying land and building houses in the country. This will be the beginning of the reurbanization process. I believe that in most cases, wealthier and more educated people will settle in rural areas, raising prices for land and real estate and forming new standards of individual housing. At the same time, less wealthy and educated people will be able and even forced to move to cities where prices for secondary housing, especially in outdated buildings, will continue to decline along with the substantial drop in the urban real estate prices.

It is logical to assume that the appearance of cities will become less and less homogeneous. The central part of the city will still develop and live its life, but the suburbs will gradually turn into a kind of ghetto, closer and closer to the center. Now it is almost impossible to imagine it, but the entire financial flow will go like a cable car from one point in the city center to a point far beyond it, bypassing the rest of the city.

The countryside, on the contrary, will flourish quite uniformly, leaving less and less land for free sale or rent.

There is one more argument that may confirm the above mentioned. With the further development of technologies, the basic values in the real world will gradually return to their original positions - air, water, food, shelter. Everything else will be synthesized in the virtual space or will be completely meaningless.

An intermediate step towards moving away from traditional offices will be to further develop the sharing model. As we have discussed, as we move to remote working, offices will begin to shift to residential areas, evenly distributed across cities. By the way, such offices are likely to be built and even leased out. In other words, several organizations will be able to pay for the construction (rent) of an office building and office equipment, and then share them with a flexible distribution of the maximum possible number of jobs assigned to each of them.

Standard workplace equipment will consist of a seat, connected to the virtual world of communications and equipped with a certain noise insulation. The perfect solution would be to develop special soundproof capsules that could allow the seat to be adjusted horizontally or vertically. People will get used to working with more concentration and comfort.

These capsules will be equipped with special audio and video systems capable of sending and receiving a 3D model of the person inside, and then, if necessary, finalize the resulting image adding some extra options in the form of suits, accessories and other things to enter the common virtual spaces. All participants of such shared virtual spaces will be able to clearly see each other, first on monitors, and then through TV glasses (which we will discuss in the next chapter). The magic is they can alter their appearances, appropriate, for example, for business negotiations. In the future, as VR technologies develop, separate virtual worlds will be created for various types of business activities. We'll talk about it later.

Moving from global trends to more down-to-earth issues, one can start with the statement that in another 20 years all kinds of real estate agencies in their current format will disappear from the market. Instead, it will be enough to use an appropriate mobile application with full access to the necessary information to buy or sell real estate. The whole sale process will be completely predictable and safe: cash will disappear from the turnover, financial transactions will be protected and insured. There will also be changes in the way real estate information is presented, considering the use of gradually developing virtual reality technologies.

In rural areas, as is already evident, over the next few decades the main demand is expected to be for small new settlements, fully equipped with all the required infrastructure. Standard farms will become popular. In addition to housing, they will contain certain infrastructure for farming. Such farms will form clusters for joint use of automated systems and robotic equipment. At the same time, ordinary houses and farms will be fully provided with their own "green" energy generators. The land will become expensive, and its price will directly depend on the potential for energy production in any form. Automated systems will easily calculate all the necessary equivalents.

In large cities, the demand for standard apartments in standard multi-storied houses will prevail, as they will become much cheaper due to new construction technologies and equipped with all innovations. The construction of new-type residential quarters will be delayed and may take place after the period under review. Too much has been built in recent decades.

Despite the fact that wheeled houses in the ordinary sense are not considered to be real estate, we will discuss them in the same chapter. Massive launch of self-driving vehicles will result in robotic mobile home market. New technologies will give people complete freedom to choose where to live, enabling to work remotely and travel simultaneously. Provided there is no need to drive a house on wheels, the development of more spacious and even super spacious motorhomes will follow. This type of housing will most like be more popular among people who are not saddled with children, but who know how society and its attitude towards children will change? Once wireless energy is available, we may have fully robotic transformer houses that take care of all our needs.

For obvious reasons, the real estate market cannot change as fast as other industries. Huge amounts of money have been invested in buildings for decades, and these buildings will be there for a long time, reminding us of the past. In 20-30 years, the changes we've talked about will only begin to manifest themselves. They should get the maximum boost at around the same time as the massive introduction of wireless electricity in the streets.

Let's now turn to the issue of construction. In 20-30 years, it should become utmost typical. This will make the technology cheaper and reduce the construction time. In general, we should have learnt long ago how to build a single-family house within a few days, and an apartment house - within a few weeks. Only architecturally sophisticated construction sites may require more time. At the same time, the construction should be typical, first of all, in terms of having a single set of components for the fast assembly of buildings and structures, as from a set. This set of components will allow to change the appearance of the ready object to some extent.

New technologies will be actively applied to the construction of detached houses. Quite soon companies will start buying land in relatively remote locations and establish the production of standard components, and then the construction of the first new-type settlements. As already mentioned, these settlements will have a fundamentally new infrastructure. They will be designed for people who always work in remote areas. The launch of self-driving vehicles will make the rest of the world easily accessible to the residents of such settlements.

On the one hand, it is quite logical that the countries with more developed infrastructure will be the first to create a new type of real estate. On the other hand, at some point, this process may accelerate in countries with lagging infrastructure, if this lag is somehow compensated for by appropriate changes in legislation.

Once new construction technologies have been finalized in individual housing projects, they will be used in high-rise buildings and structures. During production of elementary components, all necessary elements will be built into the components before installation, including usual drainage system, ventilation, home automation system called "smart home", and solutions that may later affect the external and internal design of the building. When completed, you can change the color scheme, quickly embed and add new constructions both inside and outside, and in some cases even change the number of floors.

I am pretty sure that 3D printing as the major future technology for manufacturing components will provide enough speed for large-scale production in the required quantities. As you know, this technology is unique due to its fundamental ability to create any object, no matter how complex it may be in terms of its internal structure. Thus, it will be possible to assemble structures like a Lego set without using additional binders. This, in turn, will mean that it is possible to reuse elementary components. Given the need for high speed, the construction process will be clearly divided into independent stages and will be as automated and robotic as possible.

The process of ordering and choosing real estate will remind us of buying a modern car. At the design stage, the customer will provide the basic requirements for ordering the necessary components and installation work. These requirements will include the layout, color scheme, interior design options, type of furniture, household appliances, etc. The level of detail will be down to the names of specific products. This will be very simple and very fast, based on VR technology.

3D printing technology development will enable us to build large objects entirely from different material mixtures, together with interior fittings (furniture and equipment). It is hard to say what will be easier in the end: to assemble the building from components on site or to print it as a whole. Already now, large industrial objects are built using 3D printing. In the end, the choice will be determined by the total price. In any case, when implementing "turnkey" projects, customers will be able to create almost any architectural projects (including both external and internal) that meet the requirements of constructive safety, and prepare homes for moving in, complete with toothbrushes on the bathroom sink as a compliment.

Household Appliances and Electronics

Let's discuss what inventions may appear in the world of household appliances and electronics in another 20-30 years. No secret that we are gradually moving towards the Internet World of Things, when absolutely any household appliance (home or professional) will have access to the World Wide Web by default. New-generation devices are already equipped with this option. However, we still do not fully understand how to use it, and devices of different types do not yet provide each other with information. Our transition to the Internet World of Things will be over when the information received from all everyday devices not only accumulates somewhere in one place, but each time is enriched. Besides, the devices will provide us with dynamic access to this information when we switch from one device to another.

What inventions may appear in the world of home appliances and electronics in another 20-30 years? Let's start with the already widespread smart home technology. After we start building new types of homes with all the necessary infrastructure and switch to wireless electricity, the range of customizable smart home functions will expand. In fact, it will be a full-fledged home helper that possesses comprehensive information about all processes taking place within its walls and even within a short distance of the house. We will be able to communicate with it both by voice and through standard interfaces. A smart home will know who is inside (a person whom the house knows/a stranger) and his or her exact location, where the furniture is, what condition the basic infrastructure is in, where all appliances are located and in what operation mode, and will also have access to information stored on all electronic devices.

What will be the fundamental difference between the new fridge and the one we have now? It will be able to predict what goods to order and when, to meet the family's needs, and place an automatic order for delivery from an online supermarket. The owner will only need to accept the delivery and Put away the groceries. Certainly, it will be possible to program the fridge, set the basic quantity of products, define some functions and presets. In particular, if necessary, the fridge can be used for diet control. The easiest way for the fridge will be to deny access to food at the wrong time and give cooking and nutrition tips. By the way, if you talk about how the fridge will know that it is time to order a certain product, you may need a built-in barcode scanner. However, in 20 years we expect better alternatives.

If we further develop the idea of a smart fridge, we should say that the world as a whole has long been ready to create a full-fledged food complex. The existing technologies already allow us to cook restaurant quality food at home. Ideally, a refrigerator, food storage containers, mixers and blenders, a stove (or whatever may replace it) and a dishwasher should be combined into a unit capable of preparing any dish, provided that the right ingredients are available. Naturally, this combined unit will be controlled by a smart house so that the dishes are cooked at a given time and served in portions automatically. This ensures that children and elderly people will be provided with food in no time, even without a personal appearance. In the next chapter, we will get back to the subject of our combined unit and add the missing element.

Unfortunately, the current infrastructure of developing and most developed countries still does not allow automatic recycling, and this is unlikely to happen in the next 20-30 years in old urban areas. However, new areas will be designed to automatically separate and recycle waste from apartments or other premises in order to fully control recycling processes. By the way, the first implementations of this approach already exist nowadays.

What will happen to TV in 20-30 years? Very soon we will no longer associate this word with television. TVs themselves will work differently and provide us with quick and easy access to all the information available both online and at home. In fact, this has already been happening in recent years, but so far TV control does not seem so easy. Moreover, no matter how much they try to make TV screens bigger, people still don not feel like they are in a virtual reality. The built-in sound system is not enough to achieve a completely satisfactory result, so you will have to install additional sound equipment in the room or use headphones.

Obviously, in the near future we will see new generation TVs, which will be much easier to use. Cameras and microphones will become mandatory attributes and will enable to actively use TVs for distance learning, work or just communication. And, of course, the smart home technology has been trying to use this home appliance as a basic dashboard for several years.

However, we need another way to interact with video content. We need a new device that will simultaneously allow us to fully immerse ourselves in the picture and sound, and work with both virtual and augmented reality. Certainly, the most logical solution would be to further improve the augmented reality glasses, which currently seem a little cumbersome and do not offer full immersion in the virtual environment. The new glasses will be quite thin and light, and their temples can be used as headphones, providing first-class sound and superior noise reduction. They will also consider possible visual impairments, automatically adjusting to the user's needs, enabling all people to wear them constantly.

When manufacturers implement all the above-mentioned functions in one device (let's call it TV-glasses), we will finally move into virtual reality. It's easy to imagine what opportunities people will have, not only when watching any video content, but also when communicating with each other normally. Foreign language ignorance will never again be an obstacle to communication. Glasses will be able to suppress the sound coming from the interlocutor, simultaneously translating the speech into the desired language.

TV glasses will transform video shooting. After all, TV glasses will not only show the visual imaginary created by the cameraman and director at the moment. Any viewer can zoom in or out, choose the desired angle, switch between cameras. And if there is a sufficient quantity of cameras in a broadcast hall, he feels like he is actually inside, for example, a concert hall and may turn his head like a real viewer. We will discuss this topic in more detail in the "Entertainment" chapter.

Will the washing machine of the future hand over the freshly ironed clothes to people, or instead the infrastructure will require centralized dry-cleaning services? Will robotic vacuum cleaners of the future not only vacuum and clean floors, but also shake dust out of their own containers? Answers to these questions will not be so important when people have personal robot assistants.

Personal robot assistants will become the most significant addition to the world of home appliances for the next 20-30 years. We will get the first prototypes within 10 years, and in 20 years most households in developed countries will have their own robot that is always eager to help around the house. There will also be robots-secretaries, but they will not be so common, because the work with information will still take place mostly in virtual reality.

What will the assistant robot be really like? In 10-15 years, in many wealthy families, newborns will see, from the first days of their lives, a robotic nanny next to them, far superior to human nanny. She will be able to speak any language or tone of voice, show the built-in video, access any information along with the ability to control all major indicators (temperature, pulse, blood pressure, etc.), send video with the current situation to parents, show the child video with their parents, feed, soothe, etc. The most difficult function is probably diaper changing, but there are several options: from sending an alarm to Mom or Dad to special technology solutions for automatic diaper changing. Or maybe they'll finally come up with something to change a diaper - who knows what eco activists will think about in the coming years! In any case, there is no need to change diapers very often, and perhaps, parents can live for 2-3 years until the need for diapers disappears.

Gradually, children will get used to the presence of a robot near them and will treat it the same way as today's live nannies. Over time, many of them will notice that the robot can provide much more attention and care and much more useful information than parents. The robot will not be in a hurry and will never experience negative emotions. Instead, it will perfectly mimic positive emotions. As the child grows up, technology will develop, some details will be replaced, the robot software will change, but for the baby it will remain the same - it will have the same name and will remember the whole history of its interaction with the baby from the first day of life. It will make the robot the most valuable companion of the baby for the rest of his life.

Now it is difficult to say for certain, but I can assume that children who spend more time with robots, not with their parents, will also inherit from the latter some emotional habits. Most likely, children's emotions in this case will be positively neutral, i.e., the same as those of a robot.

I began with a story about childhood, because the impact of robots on children in the near future will be much greater than what will happen at the same time when adults meet their first helper robots. This is because adults will perceive the robot mainly as a machine - maybe useful and convenient, but still a machine that should do only what is required.

Children will begin to subconsciously endow robots with human qualities and perceive them as people in a different form, trying to build full emotional relationships with them, as with living people, while robots equipped with powerful adaptive capabilities will begin to respond to children's emotional manifestations. This will gradually teach adults to treat robots as real family members.

Now, it is difficult to say if one robot will be able to combine the functions of a nanny with those of a domestic assistant or even a work assistant. Technologically, of course, it will be possible, but it is not yet clear what would be a more convenient way to use such robots for families with children. Most likely, this will depend on financial resources. The first robots will be very expensive, but very soon more and more companies will enter the market. Technologies will develop as quickly as cell phones, and over time there will be a huge variety of models at various prices.

As I have already mentioned, the most important feature of a robot of any generation will be its obvious ability to store all data during the whole period of interaction with its owners. You will never need to reconfigure a new robot. From the point of view of its owner, everything will look as if the old robot has grown or put on a new suit, while retaining its essence. The fact is that it will be able to preserve memories of the past and develop based on such memories.

The ideal robot-assistant will be able to move quickly enough, even up and down the stairs, to perform fairly complex manipulations using their upper limbs, so that children and older people will be able to lean on it. As for its "hardware", then over time it will increasingly be determined by the development of appropriate software.

Manufacturers will gradually learn to provide quick replacement of individual components in case of any malfunction caused by technological innovations or simply because the family is able to pay for additional functionality.

Most importantly, the assistant robots will also have a wide range of equipment to control all major parameters of the human body. It will not only monitor the human health, but also fully control his or her emotional state and anticipate some behavioral reactions. It should be understood, that such options will gradually put robots in some superior position and may even determine the further course of evolution, but this will not happen in another 20-30 years. So, people will still have time to think about it. As we gradually migrate into virtual reality, robotic assistants will begin to play a very important role in our lives, taking care of all our needs.

Finally, what is the future of laptops and computers? Which devices will office workers be using for their daily tasks in 20-30 years' time? I believe it will be TV glasses.

Agriculture and Food Industry

No matter how the world changes in the upcoming years, agricultural production will remain the main source of food for people for a long time. Meanwhile, today, in the age of technology, some developing countries are experiencing lack of basic food products for the population. However, in many developed countries we can observe a real "eating cult", when people consume much more food than the body needs. Although modern research shows the benefit of some caloric restriction for a living organism, the food culture will not develop in this direction. Developing countries will undoubtedly have to pass through a period of abundance of food before they start thinking about reducing calories.

In this regard, the agricultural sector will be still striving to improve efficiency in order to provide food to a growing population with increasing appetite. Solving this problem is almost impossible in today's environment without cheaper production processes and reducing the production cost of the final product. At present, agricultural companies are actively experimenting with various fertilizers and automating production processes to the maximum extent possible. This trend will continue, because good land is limited and the land we have is being depleted (let us leave global climate change aside).

All large and medium agricultural enterprises will be fully autonomous in 20-30 years. At the same time, their products will be intended for the general public and contain clear traces of chemicals, but neither the government nor people with medium and low incomes will have a choice. Hunger will be an alternative. These days we hear a lot about attempts to change people by deliberately feeding them modified food. In fact, this is a common result of increased efficiency. Business should not be unprofitable - it is against its nature. The government should have enough food to feed its population.

Therefore, any means will be used to meet these two challenges. Many governments will make the necessary legislative changes (if necessary) so as not to leave people even without modified food.

Small farms will most likely work in semi-autonomous mode, producing food of much better quality. However, they will be much more expensive, and only well-off households will be able to afford it.

It makes no sense to describe in detail automation of agricultural enterprises. In general, we already know about robotic equipment capable of providing autonomy at almost all stages of agricultural production. With unmanned vehicles and wireless green energy, we will get a full cycle in which the human presence will be necessary only in case of total system failures. Information technology will combine all these elements into one ecosystem.

Automating agricultural enterprises will gradually displace people from rural areas who do not have sufficient funds or willingness to cultivate their land through modern technologies and techniques. Since land prices will start to rise, moving to the city will be a real alternative and opportunity.

Food industry continues the agrarian sector; and here we will see the progress of existing trends. Similarly, lack of potential for significant growth in turnover will push food companies to look for ways to reduce costs and maximize profits. For some time, all kinds of automation and robotization will save them from loss of profit. However, apart from it, another perfect way of achieving savings can be cheaper raw materials with food additives, changing the body's satiation and taste quality of the output in the right way.

Thus, partially modified products (raw materials) of agricultural enterprises will be further conditioned by an even greater number of artificial additives. Previously, we only needed to increase the shelf life or give them a certain shape. Over time, it will become even more desirable to reduce their cost and increase the output of finished goods.

Considering agriculture features mentioned above, a difficult economic situation in this industry and the constant threat of food shortages for future generations, the synthesis of artificial food has long been on the agenda. We are currently aware of a number of scientific techniques in this area, which are also applied in practice. In 20-30 years, these approaches will be widespread, and the food industry will gradually replace agriculture (at least in the livestock sector).

This brings us closer to 3D printing technology. In the "Household appliances and electronics" chapter we have already discussed the next-generation combined kitchen appliances, capable of cooking any dish by themselves. We can assume that this combined appliance will use 3D printing instead of a stove. In this case, common supermarket products will not be used. For automatic cooking, this appliance will order special raw materials in the form of cartridges containing all the necessary ingredients from the supermarkets of the future. Such cartridges will be filled with products from food industry factories. Some of the cartridge contents will be synthesized artificially, while others will be natural or semi-natural products processed in a certain way to be used as combined cooking unit ink.

Obtaining original raw materials for the printer using only natural ingredients will be possible, but much more expensive. At the same time, artificial products will somewhat benefit as regards taste perception through the use of special additives, as is currently the case.

Of course, 3D printing technology will make dishes very attractive and diverse. And that will add popularity to this type of dishes. People will take an aesthetic pleasure in the food appearance, its smell and taste, no longer caring about the fact that not a single gram of natural products are used to prepare all this.

In conclusion to this chapter, we will look further ahead and try to foresee what will happen in another 30 years or more. When human chipping gets a significant scale and new generation chips are implanted directly into the brain, people will spend most of their lives in the virtual world. In this case, it is quite possible that a very simple (maybe even almost natural) but sufficient food to support life activity could be transformed into anything by processing it in virtual space and replacing the sense of taste, smell and saturation at the level of brain activity. This would provide a large population with food at minimal cost, but we would prefer to stop there without developing this idea any further, because it is too unrealistic for our time. Nowadays, instant soups containing only noodles and flavor enhancers are still popular.

Education

In 2020, most governments around the world closed educational institutions due to the spread of the COVID-19 pandemic. This fact will eventually drive global change in the entire education system. In a few years we will not have to go to another country to get a prestigious education. Moreover, very soon our children won't have to go to secondary schools and higher educational institutions necessarily. And in 20-30 years distance education will become the main standard everywhere.

When all educational institutions faced the need to switch to a remote education mode, it turned out that someone was better prepared for it, and someone worse. Some institutions did not have any experience in distance learning at all and tried to rebuild on the move to keep up the learning process. Many older teachers had to urgently master the technologies of the virtual world of communications, because they had hardly been involved in their daily lives before.

However, some institutions have long developed distance learning format alongside the traditional approach. The need for this format has gradually increased in recent years, although not so rapidly. Switching to a remote format was practically unnoticed for them.

Upon completing the forced experiment of transferring all schoolchildren and students to distance learning, many parents will breathe a deep sigh of relief as they can safely get back to their work and other activities on weekdays. However, some people who have managed to plunge deeper into the online education world, realize its great potential. Students will find great opportunities to expand their sphere of interest and realize that distance education provides more freedom. As for adults, they will know that it is really possible to enjoy the learning process.

Educational portals were previously used as auxiliary sources of information for additional education. Their creators may have already realized that in the very near future they will experience rapid development and struggle for new markets. Certainly, for another couple of years most people will think that education can be obtained only through the physical presence in the place of education. If students and schoolchildren want to take advantage of distance learning right now, someone will have to provide this opportunity to the full extent: from taking all examinations to obtaining the necessary degrees. Yesterday's educational portals will gradually transform into powerful, modern educational centers of a new generation.

Private secondary and higher educational institutions will be the next to struggle for new markets. Their owners will get the trends of the coming years and start adapting their processes. Public educational institutions will lag behind, taking advantage of the major market share, and will switch to new formats with a delay.

One way or another, all educational institutions will soon focus on technical re-equipment of their classrooms in order to establish new mandatory standards. Any schoolchild or student who misses classes for whatever reason will be able to remotely attend all lectures and actively interact with the teacher if necessary. Any teacher who is absent from work for whatever reason can give a lecture or run practical classes from anywhere in the world.

In addition to the standard communication equipment, to which we are already accustomed, the whole training process will be connected to unified automated systems. Some educational institutions will create their own IT platforms for this purpose, implementing unique solutions and attracting the best students as well as increasing the cost of education. Others will use easily accessible products offered by third-party IT companies.

New approaches will change the financial models of educational institutions. Investments in automation will pay off through the ability to attract teachers and students from around the world, which in turn will finally approve the distance education model as a target.

Distance learning will be characterized by its ubiquity and continuity. It will be available from anywhere and at any time, enabling everyone with access to the virtual world of communications to learn without limits.

But will a child from a poor family living in a Third World country be able to access educational modules remotely and go to Harvard, for example? The paradox is that they can, but will not want to, because they will probably prefer a completely different life. The educational process requires discipline, and distance learning requires self-discipline.

Most people get an education because they attend classes regularly and take exams according to teachers' recommendations. Will the child mentioned above be able to show strength of character and devote all his or her free time to self-education within a few years? Sometimes yes, but more often not. However, this applies not only to children from poor families and Third World countries, but to everyone else. This means that the educational system of the future will be strongly stratified, and instead of higher and secondary education there will be new levels - from elementary (arithmetic, reading, ability to interact with the virtual world of communications) to ultra-high (development of artificial intelligence).

In 20-30 years very few people with obvious talent and interest will receive a fundamental technical education. Only they will be able to regularly attend online lectures, discussing theoretical aspects in the same format as they do now (teacher, board, chalk).

In non-technical areas, a visual form of information presentation will prevail, allowing us to clearly visualize both theoretical and practical materials.

Considering that even today higher education degrees do not always imply real knowledge or skills of its holder. Obviously, in the future, the need for documents confirming academic background will gradually disappear. Employers can effortlessly test new candidates remotely and select suitable ones who are fit for the job.

Any self-respecting educational institution will aim at providing the best direction for intellectual development. In some cases, that's been happening for a while. But automated systems will make a real breakthrough in this area and provide very accurate diagnostic of the talents and skills of each student, so that by the end of the learning process, the transition to the employer (wherever he is) will be based on the conscious decision of all three sides.

Will children be motivated to learn the multiplication table and foreign languages in 20-30 years? Given the changes in the education system over the past 20 years and the development of information technologies, people are entrusting memory and thinking functions to the information systems.

Will young people be able to write by hand in 20-30 years? It seems pretty doubtful. To draw - yes, but to write - no longer. Digital technologies are replacing handwritten communication. Moreover, due to the speech recognition technologies and other human-machine interfaces, even typing by hand will be slow, laborious and old fashioned. The only inconvenience today is the need to say the text out loud so that the microphone picks up your speech and converts it into the text. However, with direct transmission of articulation signals, there will be no need to speak out loud. Grammar and literacy skills and concepts will be radically revised. Ultimately, information systems will be responsible for spelling and punctuation.

The transition to virtual reality will make education a very exciting process. If schoolchildren and students will have TV-glasses and specially designed curriculum (learning worlds of the virtual reality), then the humanities and primary sciences will be easily assimilated. However, the ability to memorize information in a traditional way can gradually decline. On the one hand, this may alert us. But, perhaps, humanity is gradually finding a new way to learn about the world around us.

It should be noted that the transition to distance learning in kindergartens and elementary schools will likely be much slower compared to secondary schools and higher educational institutions. Firstly, people should accept inevitable and complete reorganization of the educational system, and then it will be possible to cascade from top to bottom - from the older to the younger generation of students.

Healthcare

Nowadays doctors spend a considerable amount of time on entering all the necessary data into the computer, and it feels like they have no time for medical care. This looks quite old-fashioned, given that many doctors have not mastered touch typing. Obviously, electronic medical records, unified systems, notifications of future visits and some other improvements are already a huge progress.

If you consider healthcare in different countries in general, you will see that it is common to wait days, weeks, and sometimes months to see a doctor. This also applies to paid medical service. In general, it seems that waiting is the main "time-waster" in modern medicine. In many cases, such waiting can cost someone not only health, but sometimes life. However, in 20-30 years we will be able to use completely different health services.

The main obstacle to the development of these services, both in healthcare and in other areas, is the fear to lose jobs after the expected improvement and automation. Without this fear, the progress would be much faster. Thus, to defeat fear, people should understand what new professions will appear in the future and develop new skills. In any case, it is much easier for a mature professional to slow down progress where possible, instead of trying to improve his skills and master fundamentally new technologies and approaches. This can be easily demonstrated with the following example: some people fail to recognize the importance of touch typing, even if it can save up to 20% of their time. Moreover, neither the owners of modern clinics, nor the doctors themselves have this understanding.

Usually our visit to the doctor either ends with the prescription of treatment, or requires various tests or diagnostics, which will determine the decision on treatment. Thus, in both cases, the doctor is actually an analyst who, based on his previous medical experience (which includes medical studies, advanced training and experience) makes a diagnosis and prescribes a course of treatment. We are absolutely sure that in 20-30 years artificial intelligence will perform this analysis and render it much faster and more efficient.

In practice, you will not even have to go to the clinic, if there is no need in tests or examinations. All you need is to talk to a robotic system, which will ask you all the necessary questions, make a medical decision, a diagnosis and prescribe the required tests and procedures.

Readers will, of course, remind me immediately that usually in case of viral respiratory infection, a doctor measures the temperature, listens to the chest, prescribes blood and urine tests, and sometimes may prescribe some more serious tests (such as X-rays). And this is the main reason why progress is slowed down, given that it is a great opportunity to leave things as they are, and control the entire patient flow.

In 20-30 years, the situation will change. In fact, we should understand that we are in fact hostages to doctors who are afraid of losing their value. We need to understand that it is possible to listen to the lungs and conduct a visual examination remotely (of course, there should be an appropriate task to make it technologically possible). And, secondly, at any visual examination, any doctor, if any slightest doubt, will prescribe blood and urine tests. Do you understand why this happens? The results of such tests will show if there are any dangerous changes or processes in the body.

Most of us avoid tests after visual examinations and listening to the lungs, because the test results usually come back the following morning. At the same time, the doctor sends us home, either immediately prescribes antibiotics (and we get better), or first relies on common cold medications. In the latter case, if we feel worse, we should come to see a doctor again. If we do come back and feel worse, they will send us for X-ray or other diagnostic tests.

What would we do to get rid of the physical contact with doctors? In this case, we would focus on speeding up the results of tests and diagnostic procedures to get treatment. In the vast majority of cases, we can get tested in any laboratory. As you will recall, these are the ordinary nurses who usually carry out some basic tests. And diagnostic procedures can be avoided in the vast majority of cases.

Therefore, now we shall ensure fast and timely collection of clinical analyses and, if there are grounds, the necessary diagnostic procedures and testing of samples for further diagnosis. Certainly, listening to the chest with a stethoscope is a cheaper procedure compared with a CT scan or X-ray. But if you consider the cost of the doctor's visit, the prices are equalized. The result of a CT scan or X-ray is in any case much more useful in the long run than relying on subjective perceptions about the lung health just by listening. Moreover, even a cell phone or other portable device can listen to the lungs. It may also have other diagnostic capabilities. All examinations based on subjective perception can and shall be automated.

We all know that any disease that cannot be cured over time (such as runny nose) corresponds to a very clear and specific list of necessary tests and examinations. Therefore, we are sure that in 20-30 years, health care automation and robotization will enable to significantly reduce the time from the first appointment with a doctor to the recovery of the patient. However, health care can remain free and accessible to the public only if there are fundamental changes in the state itself. Otherwise, we will be dealing with private providers of medical services, which are able to quickly transform and automate their processes.

In practice, a patient with a potential health problem will contact his or her automated therapist (essentially a robot) connected to the appropriate information sources. After that, the robot will ask essential questions, remotely conduct an initial examination, take temperature and pulse (the next generation smartphone will be enough for this) and, if required, can direct the patient to a stationary laboratory or send a mobile laboratory to the patient to perform the necessary manipulations and take tests. The laboratory may be fully autonomous or contain a medical person (who knows both medicine and technology) to provide overall control over the patient's interaction with the equipment. In any case, the diagnostic procedure will be very fast, as the equipment will be more modern, and any delays with documentation, payment, etc. will be minimized. Unfortunately, today we still have to spend time waiting, filling out documents and making payments in the laboratory.

Then the treatment will be prescribed and the necessary drugs will be delivered to the patient's home. As for patients who come to doctors simply because they are looking for attention, they will be able to talk to an automated manager-therapist, who can no longer be distinguished from a real person. The time of such a robot will cost nothing, but it will be able to discuss common issues, remember the patient's medical records, listen to them and encourage them.

Nowadays there are extremely complex procedures, which are carried out by machines without the physical involvement of doctors. A robotic laboratory for sampling and diagnostic procedures will take blood samples from the vein easily and, most importantly, more efficiently because it will find the vein at the first attempt, observing the patient's pupils, pulse, blood pressure and possibly something else. As for STDs tests, most of us would be happier if they were carried out by a robot rather than by a human. Even now, the robot can perform any manipulation more safely and efficiently than the most experienced nurse. In 20-30 years, we will definitely have enough robotic laboratories, at least in developed countries.

Surely, everything written above refers only to the cases that do not require an urgent hospitalization, but constitute the vast majority of cases. If an automated manager-therapist has the slightest suspicion of a real threat to the patient's life and health, he can call an ambulance (unmanned, of course) immediately and send the patient to hospital. In this case, we will probably need human physicians, unless we can further automate the treatment process.

We will feel the greatest need for super professional doctors even at the stage of developing the necessary knowledge bases, setting up artificial intelligence to work with them, designing and customizing robots for operations. The functions of the next generation physician will be to understand not only human physiology, but also the basic approaches to automation. The focus will be shifted, as there will be no need to spend time studying drug formulas or special terminology in universities. The main task will be to develop methods of communication between modern technologies and medicine.

Certainly, healthcare, like many other sectors in the near future, will be gradually differentiated, following the relevant processes in society. The services available to more wealthy people will be primarily aimed at preventive health care and maximum extension of life. Artificial intelligence and robotization will make a fundamental breakthrough in genetic engineering and help control certain changes in the human body. In 20-30 years, we will be able to prolong human life by changing genetic structures and replacing some "worn-out" parts of the body with new ones synthesized in special laboratories. However, only more affluent people will have access to such services. We will not predict how much life expectancy will increase, but it is clear that the duration of its active part can be extended to a maximum, i.e. until death. This is not only about healthcare, but also about other technologies that will allow anyone to stay active until the very end.

What will be fundamentally new in healthcare in 20-30 years, which may seem surprising to us now? As we have already said, replacing any internal human organ with a new, artificially synthesized one will be quite common. As for the limbs, with the loss of arms or legs many people are more likely to choose a new generation of bionic prostheses than the synthesis of identical natural limbs in a laboratory. The main reason for this will be the enhanced capabilities of such prostheses and a certain fashion impact, which will certainly make them one of the fastest growing sectors. Bionic prostheses will not only function as lost limbs. They will be equipped with electronics, which will offer a huge range of additional functions. Already today, bionic prostheses are controlled by the human mind, which means that all electronics built into them will be controlled equally. In other words, new generation bionic prostheses can contain all the functions of a modern cell phone and even more.

If we compare bionic prosthesis with any artificially synthesized internal organ, there is no big difference between them - it is a part of our organism produced by some manufacturer. And from this point of view, we can reasonably expect that such organs should be provided with a certain interface for interaction after transplantation. External organs will be electronically familiar to us, while internal organs will require special biochips with no negative impact on the human body, but will also provide information exchange.

Technologically, we have long been ready to place various devices inside the human body. However, there have not yet been enough reasons for the serious development of this direction. Now, when modern medicine starts actively using bioprosthetics and replacing internal organs with artificial ones, it can turn to the issue of remote monitoring of the state of human vital organs. This will be an excellent occasion for gradual transition to mass chipping of the population.

However, instead of chipping all of our vital organs or waiting for us to gradually replace them with artificial ones, we can learn to extract the necessary information from our brain and transfer it to medical centers for further processing. Of course, biochips embedded in the brain will take some time to appear, allowing us not only to collect information about the state of individual organs and systems of the body, but also to provide the functionality of human interaction with the world of virtual communications directly, without using smartphones or other devices.

Such biochips will not be embedded surgically, but by means of cultivation technology with subsequent training of the brain to use them with the help of certain simulators. The chips themselves will only be used as a means of communication, and the rest can be done by our brain when it realizes which way it needs to evolve. All information about our body is already present in our heads, but we lack the skills to consciously identify and use it in everyday life. Chips will help our brain learn this procedure. Thus, concern for the life and health of citizens is the most logical reason for chipping the population, and we will consider some of the possibilities arising from this in the relevant chapters below.

We would like to consider separately our organs of sight and hearing. The moment has come when people who have lost their sight or hearing should have a real chance to get them back. And not just return, but get fundamentally different opportunities that all other people will envy. There is not much time left to get a complete replacement for our physical eye. At the turn of the 20–30-year period under consideration, new-generation artificial eyes will not only see well – but they will also have tremendous opportunities to adjust to the environment and the desires of their owner. Such eyes will see both day and night, absolutely not afraid of the sun, change focus if necessary, performing the functions of binoculars.

It will also be possible to fully restore functional hearing at a higher level. It is interesting to expand our hearing capabilities so that we can hear electromagnetic signals from sound synthesis devices without direct contact with them. It would be very useful not to use headphones, but at the same time to receive the whole range of sounds directly to the ears remotely. A new generation of people will easily agree to various experiments in this field, if they suddenly face the challenge of hearing or vision loss.

In cosmetology, total body transformation according to individual preferences is likely to continue to be a big hit. Although I think it would be more reasonable to use the possibilities of the virtual world to form the necessary images, now it is difficult to predict what will develop faster. These images may first be created in the virtual world, and then, after the medical technology has reached the necessary level, people will transfer their virtual fantasies into the real world, and then both technologies will combine and develop a completely new mechanism of body transformation.

When we move to the era of wireless electricity, all diagnostic medical devices will be contactless. Wireless electricity will let you see from a distance all existing electrical circuits and also change them if necessary. It is easy to understand that in this case we will be able to receive information about the health of any person as a continuous stream. Our brain will be one of such circuits. Maybe the technology behind wireless electricity will eventually give us a new generation of EEG devices. These devices, in turn, will show how the brain works. Maybe then we will be able to create a real artificial intelligence that is as close as possible to what we have at our disposal today.

Sports and Tourism

In 20-30 years, in the world of sports there will be two completely new branches: esports and virtual sports. Esports will start developing earlier than virtual sports as we will not have full VR technologies so fast. Virtual reality in the full understanding of this term should still include not only the substitution of audio and video channels, but also a tactile channel, and this will be possible only after connecting the virtual world of communications directly to the brain.

Esports will actually continue the modern Paralympic movement, where yesterday's disabled people will start using bioprostheses of new generation and with their help will easily surpass the speed, strength and endurance of ordinary athletes. There will be a moment when people can reveal a new potential of their body. Imagine a runner, who was chained to a wheelchair yesterday and today can run 100 kilometers in 3-4 hours. Imagine a swimmer who crosses a 10-kilometer distance in an hour between neighboring islands. Realizing the potential of a new generation of prostheses, even some healthy people will consciously start to replace limbs to get the same freedom of movement.

The intermediate stage in esports for ordinary people with no physical disabilities will be the use of special exoskeletons, because parting with healthy limbs will seem unusual for many. However, with the emergence of new generation bioprostheses on the market, there will be a shift in public opinion. After all, even in our time people do unfathomable things to the body. In addition, bioprostheses will become interchangeable, like razor blades, and maybe even leased out - who knows?

Most likely, esports will be of greater interest to the older population, who are used to being more physically active. Obviously, the heart will need less effort if the limbs work with alternative energy sources. And this will allow even older people to discover their youth and freedom of movement.

Virtual sports will be more popular among young people. The rate at which this trend is spreading, of course, will fully depend on the development of VR technologies, as we have already mentioned. The essence of this trend lies in imposing virtual reality information layer on traditional sports. It will begin with workouts that may open up new opportunities for amateurs and professionals. A new generation of simulators used in combination with VR technologies will turn monotonous exercises into a fascinating process. Running on a treadmill will turn into jogging on the seashore or a race, provided that there are others willing to take part. An exercise bike will enable visiting the most exciting places that modern cyclists can only dream about. The machines themselves will be equipped with additional devices to simulate different types of terrain and synchronize with the information layer in virtual reality.

The next step will be to provide sport facilities for paired or group sports with special video equipment, motion sensors and other electronics that will organize matches with a virtual opponent or players located in other cities or countries.

Until the full immersion into the virtual reality becomes possible, in addition to TV glasses, such sportsmen will use special sports equipment to feel the moment of contact in the game. For example, a tennis racket should transmit the pulse when it hits a tennis ball, and in soccer, sneakers should perform this function.

Switching to virtual sports will also affect the spectators. As I have mentioned before, TV glasses will also affect the formats of fights and matches. We can talk about it later in the "Entertainment" chapter.

Following the launch of full-fledged VR technology, the usual physical professional sport will gradually disappear. In most cases the age of professional sportsmen is very limited nowadays. New generation will be more and more inclined to virtual sports and esports. In time, these two areas will merge together.

Of course, we will be able to watch the usual (natural) game of soccer for many more years, but with the above-mentioned sports tendencies it will seem less dynamic. Super-dynamics will now move to the virtual sports and esports world where speed, strength and maneuverability should reach new levels. Not only good physical shape, but also mental health will become essential qualities of an athlete. The ability to concentrate will be the most important factor. Strength and endurance will be relegated to the background as technology advances.

What will tourism be like in 20-30 years? There is a reason that we consider the subjects relating to sport and tourism in one and the same chapter. In some sense, tourism trends in the coming years will repeat sports trends. Of course, ordinary tourism will not disappear completely, but will be gradually supplemented by other forms. Esports development and new generation individual transport will result in cyber-tourism. Promoting VR technologies, of course, will lead to a virtual tourism.

New generation exoskeletons and bioprostheses will significantly expand the opportunities for hiking. In addition, due to new generation transport, anyone will be able to visit very remote and inaccessible places. And if you add unmanned cars and the Internet everywhere, even traveling around the world will not seem so expensive or difficult. Thanks to wireless electricity, we will eventually have unlimited freedom to explore the remotest places and to visit the most amazing and hidden corners of our planet. This is what cyber-tourism will be like.

Certainly, some will still prefer to come to Paris to climb the Eiffel Tower and drink champagne instead of climbing Everest or crossing the Sahara. But virtual tourism will make such a trip much faster, cheaper, and most importantly, more picturesque. Virtual tourism will be fundamentally different from the traditional. The fact is there will be no need to go anywhere physically. The virtual reality industry will give a boost to a number of companies specializing in creating virtual worlds for tourism and training guides. Anyone can visit the Eiffel Tour directly from home by connecting to the appropriate service. Tourists will benefit greatly from such virtual travels by shooting at extreme angles and immersing in breathtaking experiences. For example, ordinary (natural) tourists can never do something like jumping from the top of the Eiffel Tower.

Once we start using full-scale virtual reality technology, it will be possible to climb Everest from home, providing complete and realistic experience. But I think it would still take more than 30 years.

Hospitality

How will hotels and restaurants change in 20-30 years? Clearly, these industries will never grow as we may expect. Hospitality owners will be puzzling over how to complement their core business models with new revenue streams.

Large automation and robotization are long overdue in the hotel business. Now, of course, it is difficult to imagine self-cleaning rooms and an admin robot at the reception, but in general, technology shall develop this way. Cleaning a standard room is a rather predictable thing, repeated many times in the same routine. So, it would be reasonable to replace human labor here. However, the first fully automated hotels will have very few rooms, and they will appear closer to the end of the period under consideration.

Cleaning of the room involves change of linen, washing and cleaning surfaces, daily maid service, maintaining the equipment and minibar filling. Most likely, the automation of these functions in existing hotels will be gradual and take more time than in new ones, due to the usual infrastructure toughness. At the same time, all entrepreneurs determined to build new hotels from scratch should consider designing the infrastructure in order to ensure automatic circulation of all replenished room equipment (linen, towels, soap, etc.) through special channels built into the supporting structures, and embed in advance all electrical infrastructure required for the phased automation. But first, of course, we need to think carefully whether the construction of a new hotel will be a good investment with such major changes in consumption patterns.

Since the flow of tourists will gradually decline, some rooms in new hotel types (primarily in non-resort cities) might be equipped with full-fledged studies or personal offices with access to all modern communication technologies. Obviously, with the advent of unmanned cars, people will practically stop staying in hotels for a day or two. It will be much easier and cheaper to spend the night in a car while it is taking you to your destination. It will also be cheaper to buy a portable eco washbasin and save on accommodation or maybe use gas-station toilets, which in developed countries are almost as good as bathrooms in decent hotels. This way, most customers will stay longer.

Hotels with the highest level of automation and robotization can offer more attractive prices to convince self-driving car owners to stop and sleep on comfortable beds. However, the fight for the client will still be tough. Therefore, the number of hotels will gradually decrease, although now it is difficult to predict which segment will suffer the most. On the one hand, automation may require serious investments. So, we can assume that small players will leave the market, and medium players will gradually move to larger ones. On the other, it may turn out that big players will not be able to rebuild the infrastructure as quickly as the medium ones, so they will start selling their property and gradually lose market share. Small hotels have very little time left, and they should start planning now what to do with existing assets in 20 years.

Due to seasonality, resort hotels are in an excellent position to upgrade. In any case, regardless of outdated hotel infrastructure, automation of many obvious processes can significantly increase both customer loyalty and profit. For instance, check-in and check-out registration. Simple automation can significantly accelerate these procedures, and the client will be grateful for the time saved. In addition, most hotels do not consider room service to be a serious source of income. Although even now there are excellent opportunities to increase income from these services through the automation of orders.

However, hotels of the future, which plan to stay in the market for more than 20-30 years, will soon begin to change their approach and expand the list of opportunities for their clients. Future automated systems will enable us to plan personal schedule in advance, serve meals at a certain time according to a pre-agreed menu, provide opportunities for tourism and recreation, and if necessary, act as a realtor. And, certainly, all these services will be available in the clients' native language, wherever they come from. The future hotel will focus on making each client feel unique as if he is the only one on vacation at the moment. Of course, we are not talking about beaches or swimming pools, although some flow of clients can also be distributed depending on their preferences of each client.

Will people go to restaurants as often as now, in 20-30 years? We do not think so. As we know, restaurants provide food and communication. We have partially discussed what communication in society will be like in 20-30 years, and we remember that it will gradually move into the virtual world of communication. Even now we can often see close friends or acquaintances sitting next to each other and looking at their cell phones rather than interacting with the person they are with. Young people are communicating with each other via messengers, even while sitting at the same table. This is quite in line with the times.

This is how we lose communication. All restaurateurs might either ban smartphones during dinner (and maybe lose a group of customers, but save the atmosphere of a real restaurant), or turn waiters into food delivery service, and transfer the food ordering process to the virtual world of communication (and maybe lose another group of customers).

The most expensive and exquisite restaurants can follow the way of prohibiting smartphones for the next 10-15 years until the next generation of customers. But they will be forced out of the market, as by that time the consumption pattern will have fundamentally changed. Other restaurants need to think about the automation of restaurant premises, completely abandon the concept of waiters due to the launch of robotic analogues, and simultaneously get ready for maximum automation and robotization of the kitchen.

In the "Household appliances and electronics" chapter we have already discussed a new type of kitchen processor, which in certain configurations can prepare excellent restaurant quality dishes right at home or in the office. However, not everyone can afford such a novelty. So, people will go to restaurants just to eat quickly, remaining in the virtual world of communication.

Besides, food processors for cooking will be an excellent device for restaurants of the future, which will have access to our apartments and offices, offering unique dishes by subscription. The restaurants will own advanced models in different configurations. At the same time, TV glasses will complement our daily home or office reality with the restaurant atmosphere and allow us to communicate with colleagues and friends at breakfast, lunch or dinner.

Entertainment

In the next 20 - 30 years the show business as we know it will gradually disappear from our entertainment industry and move into the virtual world.

We are already observing the most powerful development of online cinemas. We can confidently state, that physical cinemas will disappear first, and soon enough. Their era will end with the advent of TV glasses, which we have discussed in the "Household appliances and electronics" chapter. With the appearance of TV glasses, cinema production will reach a new level. In the future, camera operator and director's work will enable viewers to be in some sense almost in the frame, near the actors, and to see everything that happens around with their own eyes. It will be important to work out the smallest details both in the frame and behind the frame - people will watch the spectacular moments many times from different angles. Acting skills will be evaluated only with regard to directing and operating, because the actors will be completely dependent on them. These are primarily the directors and camera operators who will be the first to hit the red carpet of virtual worlds. These professionals are able to turn the world inside out and show the audience the most unexpected angles. They will turn into some kind of sculptors, modeling virtual worlds with their hands. However, they will benefit from the appropriate equipment, which shoots in "3D 360 degrees" and creates absolutely accurate and realistic copies of the real world, which may later be converted with the assistance of information technology.

Physical cinemas will disappear a little bit later, because they are still closer to natural art and therefore more difficult to transfer into virtual reality. More precisely, such transfer is very sensitive to the level of shooting technology or image display. After all, to create a real illusion of being in the theater, it is necessary that the virtual viewer can feel like he is in the hall, sitting at home on the sofa. This requires not only technologies to create virtual worlds, but also technologies that simulate the presence in the virtual world. This will be achieved in 20-30 years. And again, unfortunately for actors, directing and camera operating will come to the fore, as the viewer should pick up on emotional cues, facial expressions, feelings of actors in all details, as well as be able to approach them and study every elusive detail from different angles. Plus, they can feel the whole atmosphere of the theater and the presence of other spectators.

Then theater and cinema will merge together, and the world will get the symbiosis of beautiful acting and the opportunity to enjoy the interiors of the theater and the environment in general. In fact, virtual reality (VR) technologies will offer theatre and performance unique compelling possibilities instead of physical decorations.

Only virtual presence technology will save the classic theater, where each viewer somehow finds himself in the virtual hall with the others. This can be achieved with the help of cameras built into next-generation TVs (at least, until we implant biochips in the brain). These cameras will capture our presence in the virtual theater, using virtual images that we create in advance or purchase for the event. These images will be superimposed onto the movements and facial expressions, creating a complete image to be transmitted to other participants in the performance. However, as we have already noted, the result will not be perfect without chips, and therefore theaters may simply not reach this point, absorbed by the film industry.

Like in theaters, when we go to a concert or a sports match, first of all we expect certain emotions that we cannot experience when watching remotely. We go there to bring home the reality of what is happening, for live sound, for communication with real people, for noise and fun, and for a festive atmosphere in general. We will have all this in virtual reality. New technologies will take us, physically sitting at home, in a real hall or stadium. There should be other people around us, and the atmosphere should be the same as in real life.

The main difference between a virtual concert or sports event (or theatrical production) and watching a movie is that the action takes place in real time, as a live broadcast, and musicians or athletes perform or play (compete) in real time right in front of the audience, wherever they are. To ensure and enhance the effect of presence at the venues it will be necessary to provide them with video equipment of a new generation and to develop special software allowing each viewer to see any part of the space, whether it is a stage (soccer field, tennis court, etc.) or any other area around them. All this will be absolutely real.

In this case, the viewer will be able not to just turn his head and look in different directions, but at any moment move to the point of his interest, as close as possible to the spotlight, and see what's happening at just the right time, not when the director and camera operator switch to the proper camera as they see fit, and certainly not on repeat. To be near the goalpost or behind the goalkeeper at the moment of attack and to see the ball passing by, to see the emotions of players just a meter away from you - isn't it the highest pleasure of a real fan? Or to get close to a musician or a singer, the idol of your life, and listen to them playing or singing, standing just a meter away from you?

Of course, in each case, the viewer will be able to watch a match or a concert in splendid isolation, adjusting the desired environment in accordance with his requirements, considering the options available in the corresponding video broadcasting service. This will create opportunities for additional monetization. In addition, until the technology becomes capable of simulating the virtual presence of the viewer, there will be no other option but to watch it alone.

I have examined the issues of the film industry, concerts and sports events in detail in order. And now let's consider other types of entertainment: they will also gradually move into virtual reality. Museums and exhibitions will do this first. Even the existing equipment is almost enough to achieve the desired effect. However, TV glasses will complete this process.

With the release of TV glasses and new video equipment, zoos will gradually deteriorate, as the feeling you get when visiting zoos of virtual reality will be more intense. A new generation of visitors will first get used to a bright picture, that can be controlled on their own, instead of trying to watch something through the glass or climbing over the fence. During classes, educational institutions will be able to use live broadcasts from museums, zoos and exhibitions.

Circuses, fashion shows and all other events will follow the same path as concerts and sporting events.

For the complete transition of amusement parks, fairs and clubs into the virtual reality, we need chips in the brain or another way to give us the appropriate tactile information. Otherwise, full immersion will not happen.

Services

Let's consider some popular services that have not been included in the previous chapters and see how they can transform over time.

Many people visit hairdressers not only to get their hair cut, but also to chat about the latest news, and to spend time in general. However, more and more people go there just to get a haircut. Will we be able to get a haircut in the near future - say, in 20 years - from a robot that will talk to us only at our request and only on favorite topics? Will the robot, without asking questions, simulate a haircut based on a photo or memories from our previous visit, clarify a couple of important details and then quickly do a haircut? Will it be able to offer us the right hairstyle, immediately show us how we will look from all sides, then slightly refine the image and accurately implement it? Will the robot eventually be able to come to us and do the same at home, in the office or somewhere else? Well, not a robot, but a self-driving car with a built-in automatic hairdressing salon.

How will cars and other vehicles be repaired? A self-driving tow truck will pick up your car, take it to a car workshop, where robotic systems will quickly replace faulty parts and return the car as soon as possible. As for regular maintenance or minor repairs, the vehicle itself will go to a service center at a convenient time (e.g., at night) and then return at a scheduled time. If the repair takes longer, then a replacement car will arrive along with the tow truck (now such service is already quite common).

By the way, if we get a bit off-topic here, the transition to self-driving cars, as we have already said, will be accompanied by the further development of the car-sharing model. City residents will rarely think about buying a personal self-driving car. This means that we don't have to worry about repair or maintenance. We will use the closest vehicle available, knowing that it is serviced regularly and, therefore, is fully functional.

Outside the city, having your own car will be more relevant. In any case, physical car dealerships will gradually leave the market, since the population density outside the cities is quite low, and in urban areas, there will be several large sharing dealers who buy self-driving cars. Thus, these operators are likely to perform car maintenance and repair. In addition, at some point in time, the main car spare parts will be produced directly on the territory of these operators, on 3D printers, which may eliminate the need to keep an excess stock of spare parts in warehouses. For a manufacturer, the best way to sell a car will be supplying car-sharing operators to contact their potential buyers directly while traveling.

How will dry cleaning change in 20-30 years? In due time, an autonomous car, equipped with special compartments for clothes, will drive up to our house. At the entrance, it will recognize us using a camera and (or) voice recognition system and indicate which compartment to use. After that, the car will drive away, and when it picks up all the orders, it will immediately go to the laundry, where various robotic systems sort all the clothes, clean them and contact the owner if there are any problems. Upon completing, the autonomous car will deliver clean clothes to the right place at a convenient time, automatically receiving payment from the client's account. Of course, you can track the status of your order at any time. If you have your own robot assistant, it will interact with the dry-cleaning service - send dirty clothes to the laundry and hang clean clothes in the closet.

With the decrease of retail stores, fitness clubs will occupy much larger areas in shopping centers (due to lower rental rates), adding indoor tennis courts, mini-football fields and other similar facilities (which in the future will be supplied with the necessary equipment for virtual matches). Maximum automation and robotization will make human personnel completely unnecessary. All equipment will perform self-diagnostics. Robots will do the cleaning. Security will be ensured by an intelligent video surveillance system. Fingerprint scanners will be installed in the locker rooms. All equipment will be contactless. Fitness trainers will switch to virtual reality, as their physical presence will no longer be required. Given the very negative experiences from the 2020 pandemic, all fitness business owners will first and foremost consider going completely offline and implement approaches in which individual visitors can, in fact, not come into physical contact with each other. The pandemic may never happen again, but no one wants to risk a complete shutdown and subsequent loss of business. In addition, technology itself is pushing us towards large-scale change.

Future atelier also stands a good chance of becoming fully automated and robotic. In fact, we only need a functional fitting room that will accept an order, demonstrate the result and agree it with the client. The cutting and sewing process has long been automated. The only reason for the delay was due to no solution to automate the entire business process. However, physical clothing stores will gradually disappear as the fitting process becomes fully virtual.

In general, the creation of robotic studios providing certain services can and should become one of the directions for the development of modern information technologies. Algorithmic formalization of business processes will then be used to design and, for example, print a robot on a 3D printer to perform the necessary functions. 3D printing shall ensure that the robot is equipped with all the electronics required to operate automatically.

In this case, all internal microcircuits will be printed immediately according to the draft design and will only require the installation of the necessary software. In other words, visual programming of business processes should lead to the creation of a finished prototype in a language acceptable to a 3D printer, with software for the prototype. Ideally, of course, the technology should be controlled by artificial intelligence. It will only need instructions to learn a specific business process, after which it can do everything else on its own. However, this will be possible not soon enough.

Advertising and Marketing

Nowadays the Internet sometimes knows much more about us than our relatives or friends do. When we now receive advertising messages, we begin to notice: it looks like someone knows what product or service we really need. The more information we leave about ourselves on the Internet, the more complete the picture is about our preferences. Various automated systems exchange data with each other, creating in pieces (like doing a puzzle) our psychological profile, accumulating information, which includes some payment details, addresses, phone numbers, data from documents, photos, etc. Everything that we once transmitted in any way, potentially replenishes all kinds of knowledge bases. Therefore, referring to the prospects of marketing, we can affirm that very soon specialized marketing systems will form a complete list of required goods and services for any person in the near future. Manufacturers and suppliers, however, will only fight for a certain brand name to occupy a particular position on this list. At the same time, marketing systems will, of course, know that you have been using a certain type of toothpaste for many years, and, moreover, they will know the reason for your choice and also understand what exactly can make you change this choice. And they will sell this information to manufacturers and suppliers.

Advertising in our time, as we can see, is still designed for massive views by various groups of people. Based on marketing analysis, it reaches the right addressees. Advertising is now almost never personalized. It uses more recognizable images, plays on the eternal emotions and feelings, but the content is, in fact, universal.

It all depends on advertising and marketing budgets of manufacturers and suppliers, which usually imply the maximum possible savings. Considering that at the moment the largest component of these budgets is the temporary rental payment for some conditional information space, but, eventually, we get minimally personalized advertising, which is viewed by millions of different people. The only exception, as I have already mentioned, is the Internet, where access to semi-personal (if I may say so) and sometimes personal user data has already made it possible to divide advertising flows among target groups.

Well, what will be advertising and marketing like in the era of virtual film screenings, concerts, sports matches or other events? We can imagine several million people sitting at home and waiting for the beginning of a sports match. The marketing module of the system has already received and analyzed all the necessary information (at the time of payment for virtual viewing, or even earlier - when registering in the system, providing a particular video service) - name, gender, age, social status, place of work (profession, education), approximate capacity to pay, a list of all acquisitions over the past few years, and certainly a number of other data. As I have mentioned before, this data will be collected over time by analyzing all types of sources using intelligent search engines and collecting relevant information, and then collected by any interested service for subsequent monetization.

The marketing module of the broadcasting system will easily be able to send each of those millions of people a suitable advertising proposal. Considering the preliminary detailed analysis of the audience, the cost of renting information space will be calculated differently - as has long been practiced on the Internet.

If, for example, the broadcasting system knows that the viewer has already bought a certain brand of beer before the match (the marketing module of the video service bought this information in an online supermarket), there will be absolutely no need to give such a viewer a standard advertisement of the same beer, once again annoying the consumer and thereby undermining the loyalty. If the broadcasting system understands that this viewer cannot afford a new car of a certain brand and the next loan may not be approved soon, is it worth spending the marketing budget of the car manufacturer on this buyer?

Thus, at the same time, millions of viewers will watch various advertisements. Advertising should become personalized. Should a father and son, sitting next to each other on the sofa, watch the same video with products that theoretically could fit each of them? Buyers may have different motivations for shopping. Previously, technology and budget were not enough to produce and broadcast personalized advertising. In the future, this oversight will be corrected.

If we turn to the usual formats of modern advertising on television, it is safe to say that such advertising will soon disappear. Those principles, which were relevant 20 years ago, even nowadays work more for rejecting than for increasing sales. And the demand for print advertising in all its manifestations will no longer be in demand.

We expect a real advertising boom, which will be displayed on all new generation equipment, fitted with information boards of any type. As we remember, soon any device will be connected to the Internet by default, which will make it suitable for advertising purposes. If household appliances can be reconfigured, then the situation will be completely different in public places. Given our openness to all types of social networks, as well as the ability to connect almost any device to such networks, our user profile information will be used to generate appropriate promotions literally on the go and broadcast them to those devices in public places that are currently near our location.

Consider a simple example: imagine that you are going somewhere by self-driving taxi, which obviously knows something about you. It knows enough for the car next to you (even if it belongs to a completely different company) to create a unique offer at this moment and display it on the rear or side window as you drive in traffic. Of course, no one will share specific personal data with other companies, although this may happen from time to time. However, even a set of enough general information (gender, age group, car type, time of day, travel area, weather, news context, etc.) will help to identify your potential needs to create a promotional offer for you. If we meet another taxi from our car sharing company, we will definitely get only what may be useful to us.

The road billboards will work likewise. Focusing on the flow of cars, they will change their offers, especially if the traffic density is not very high.

Logically, this transformation of marketing and advertising should eventually lead us to receive offers that are so personalized that we will gradually start paying more attention to them, knowing that we risk missing something really worthwhile if we ignore them. After all, it is very difficult to identify something meaningful for us in the vast amount of information that surrounds us in our daily life, if we do not track a large number of different sources purposefully. Only then we will be able to take advantage of what has been so annoying for years.

The Media

How will the media change in 20-30 years? What will television, radio, periodicals be like? These questions are almost rhetorical, but nevertheless, the matter is worth considering.

I have not watched public television channels for more than 15 years and was surprised to find out that TV stations still broadcast the same programs as before. No one I know watches these channels either. This is due to small amount of quality content (informative and authentic), advertising overload, and inability to fast-forward through the commercials and give feedback. Nowadays public television is used only by people who do not care about the above-mentioned circumstances or who are deprived of alternative sources of information for one reason or another. I would not say that in all the countries, including developed ones, public television is not airing informative and trustworthy content. Certainly, it depends on the maturity of the society and the state system.

From a common sense perspective, information paid out of taxpayers' pockets should contain minimum advertising (ideally none at all) and be as informative and transparent for society as possible. However, as it is very tempting, in many countries state media are more often exploited by third parties, particular individuals who are at the head of the country, and act in their interest, rather than in the interest of society.

This phenomenon brings both disadvantages and advantages. However, a modern thoughtful person cannot help but seek to obtain the most objective information, i.e., pure facts on which he can draw his own conclusions. Therefore, most of the modern audience is already actively looking for new sources of information that could be more reliable.

As we all realize, the vast majority of countries have public and commercial media. At the same time, in more developed countries commercial media may not be controlled by the state at all, but journalists are in any case paid salaries or fees, which means that sometimes they may be asked to do something, directly or indirectly. A journalist may not always be able to say "no" to his employer, highlight this or that, change the presentation of information or even manipulate facts. As we see, commercialization of all media will not solve the problem of potential deterioration of their quality.

However, today's reality is that any information channel can very quickly lose the public trust. It is enough to broadcast distorted facts a couple of times, which inquisitive viewers can easily double-check and, having discovered fraud, publish a refutation - and all the reasonable audience will be lost once and for all. People do not like to be deceived, even though they love deluding themselves. However, even the biggest admirers of delusional ideas get very angry when one day they realize that they have been deceived deliberately.

We believe that the next 20-30 years will be the era of a really independent media. How this period will end and what the media will transform into upon its completion depends on the position of society and the state. Now we finally have the technological opportunity to see an objective picture of what is happening in the world, to free facts from other people's opinions and to resist attempts to impose other people's conclusions on us.

Regarding print media, the simple fact is that in 20-30 years, we won't buy paper versions, as most newspapers will have to shift to digital online editions. Internet news aggregators will collect information, verify its authenticity using AI technologies, and then assign trust ratings to it before final publication. It will be almost impossible to get the maximum trust rating, but a high rating will indicate a sufficient level of accuracy. This approach will make the public realize that absolutely any information should always be questioned no matter how reliable it may seem at first sight.

News aggregators will enable us to access only the types of news we are interested in, which may be aired in a convenient way (text, video, audio) on those devices we set up for this purpose. Advertising, as we have already mentioned, will no longer be annoying, so that we may accept it as some kind of news: that is how it will be presented. New aggregators will also save us from the need to read news containing errors due to automated checking before publication. Nowadays, unfortunately, even the headlines of the most famous media can contain grammatical errors, which undermines the trust of readers.

As technology advances, news aggregators will become a kind of new censors. All information, whatever form it comes from different sources (audio or video), will influence the rating of sources publishing it, which will cut off unscrupulous news providers.

We have gradually realized that very soon the concept of television will be completely transformed. The audience, which can still support the ratings of today's TV channels, will readjust over time and go elsewhere. However, the channels (if they retain at least some of their former popularity) will start to change rapidly, trying to find new formats of presentations and turning into video channels already familiar to us on the Internet. They will compete with other independent and popular video channels. However, firstly, they will try to just buy them out and make them play by their own rules. Then we will know if this or that society is mature enough to resist such methods of capturing market share. In some countries, one of the obvious ways to deal with independent media will be to shut down news aggregators or video services in the information field, replacing them with domestic ones, so that the state media can regain a dominant position.

In the coming years, anyone will be able to become a truly independent reporter by making video or recording interviews at any time. There are already real strategies that aim to reward people whose information is of public interest. Accounts with the large number of page views attract advertisers and make them pay such newly-minted journalists for placing advertisements on their channels. Now the most ordinary people dictate what kind of ads they can and would like to post on their channels. After all, it is not easy to gain millions of loyal viewers, but it is very easy to lose them again, among other things, due to annoying and tasteless ads.

As for modern radio, it is certainly migrating to the Internet. Thus, due to gradually developing aggregators of audio content, listeners will be able to choose from a variety of programs. Migration will happen instantly as we move towards the Internet of Things, when all devices, including self-driving cars, will provide us with access to various audio channels.

It should also be noted, that the above-mentioned news aggregators of the future will produce both video and audio news channels that we can watch or listen to whenever we want. All other non-news content will be qualified as entertainment or educational and will be published separately.

To conclude the chapter, no matter how the media changes over the next 20-30 years, we should keep in mind that there is growing number of devices capable of seeing and hearing everything that goes on around us. These kinds of witnesses are incorruptible and, if properly handled by society, will no longer enable distorting our history.

Financial Institutions

As you may remember from the "Information technologies" chapter, we have already discussed the changing role of IT professionals in various organizations. Banks are the most obvious example of the very dynamic growth of IT. Like I said, 25 years ago, many quite famous banks could still manage small IT teams to get the required level of automation, and small banks could only have one or two system administrators. Only the largest banks started to think about it and realized a rapidly growing role of IT. Now we see that all large banks have huge IT functions with a complex structure at their disposal. CIOs are increasingly being introduced to the boards of banks, occupying positions very close to CEOs, and we will soon see that this is not the limit.

However, over the past few years we have witnessed the rapid entry of new, previously unknown banking organizations into the markets, which are rapidly gaining the market share. These organizations have become the first representatives of a new industry (now they are often called "fintech"). They were created by people directly related to IT. These organizations enabled us to look at the banking sector from a different perspective and to realize its slow development compared in the face of the rapid pace of technological progress. These organizations also prompted all major banks to think hard about changing their processes in order to retain the customers who had already started withdrawing money from their bank accounts.

Bank clients suddenly realized that money can be fast and even super-fast. Large banks, having acknowledged the risk to lag behind, began either to invest actively in upgrading their business processes or to buy fin-tech companies in order to reduce the time required for restructuring.

Surely, in the near future, the largest banks will not lose their market share completely, but some changes in the next few years may make their position very unstable. In layman's terms, banks focus on granting loans, safeguarding savings, and securing various financial transactions. These three fairly simple functions put the banking sector at the top of the value chain.

But what will happen when our money becomes electronic and protected by default? What if you do not need to move it physically, but only store data about its current owner? What if all payments become absolutely transparent and you can get refunded at any stage? What if the entire life of every digital dollar is stored in the information system without any possibility of changing it? What happens if we start using crypto currency?

In this case, any IT company can develop and implement a solution to redistribute cryptocurrencies, issued by the central bank (or directly by the state), of both individuals and legal entities, without involving any additional financial organizations. They can also charge interest, if necessary, cancel all transaction chains by returning the funds and manage loan agreements and history independently.

The banking sector is perhaps the most conservative of all for obvious reasons. However, the possibilities of the technology are more than enough to cover security requirements for managing electronic money and, in particular, crypto-currency. At the same time, all banking processes are quite easy to automate (from this point of view, we would even call them classic). However, it is extremely difficult for large banks to engage in automation, given the 20-30 years of accumulation of various, often unrelated IT systems. The transition to cryptographic currencies should logically simplify all the processes in any bank, making most of them redundant. However, the difficult legacy of the past and the need to transfer the old history to a new era will make switching to cryptocurrencies and abandoning traditional methods of calculation extremely difficult for large banks.

Meanwhile, as I have already mentioned, any IT-company without any history and accumulated problems will be able to build an industrial platform and integrate it into new governmental economic processes within just a few weeks. State banks with history will seek to retain their old customers at any cost (whose transition to crypto-currency may be delayed, which will require maintaining all existing infrastructure and do things the old-fashioned way) and attract new customers who are fully prepared to work exclusively with crypto-currency. This attempt to have a cake and eat it will cost taxpayers a lot. Many governments will probably try to help their banks by creating barriers for new companies entering the market, although not objectively necessary.

One way or another, the world will have a sufficient number of international and national fin-tech players, who will work exclusively with crypto-currencies, providing their clients with incredibly fast and diverse financial management services. Commercial banks will also gradually turn into financial technology companies, or go bankrupt, or will be taken over by state banks. The latter will continue to lag behind the market, adding the clients' assets to their portfolios. We speak only about those clients, who, for some reasons, will be ready or forced to use lower-quality services.

However, among the large state and commercial banks will be those who have already understood or will soon realize that the future belongs to information technologies and virtual communications. Having serious financial resources and reputation, they will start to insist on their own transformation into IT companies, quickly acquiring various technology companies that offer innovative approaches in various industries. Eventually, former banks will turn into IT giants, supplying goods and services to their clients while providing them with new financial services. I guess, that within 5-10 years, at least one such giant will appear in each country.

Besides banks, we would like to discuss some other financial sector organizations. In 20-30 years, insurance companies should disappear from the market, turning into an appendage of financial companies. All business processes of insurance companies, if built from scratch are not something extraordinary (taking the above-mentioned banks as an example). So, with transition to cryptocurrencies, launch of self-driving cars, development of health and safety in general, only super-fast companies with maximum process automation will be able to earn at least something in this low-margin market.

Brokerage companies, which are actually intermediaries between the final holders of financial resources and exchanges, will be under serious pressure, as well as insurance companies. Actually, only large players who invest in promising services in advance will be able to stay in the market. Timely use of VR technologies for the establishment of advanced trading interfaces will help if the exchanges or financial companies do not outrun them.

Now it is quite difficult to predict how the work of brokerage companies and exchanges will change after all market participants start using artificial intelligence. Clearly, in 20-30 years live brokers will be out of the question. But we cannot be sure how different artificial intelligence will interact with each other. Just as today, the best analytical and decision-making algorithms with large infrastructure bandwidth will most likely win.

As for the development of VR technologies, that's not hard to guess that new internal financial settlements will gradually begin to appear, as the industry of virtual world develops. At first, no legislation will regulate them, as was the case of cryptocurrencies. However, in 30 years we will not be talking about cryptocurrencies, which we are waiting for today, but about virtual cryptocurrencies, given that more and more people will find a way to make money in virtual worlds. Unable to somehow influence trading on real exchanges, young and talented people of the next generation will create large-scale corporations, but already in virtual reality.

To conclude this chapter, let's come back to some burning issues. How long do we have to wait until credit cards become completely outdated? Even now they look like a relic of the past, especially the magnetic stripe, which has not been used for years. Technologically, we have long been ready to completely abandon this payment method, but the transition to virtual cards is complicated by the fact that many still have phones that do not support the technology required for contactless payments. It seems that within the next 20 years we will get a dozen more envelopes with plastic cards until all phones are updated. It will be very interesting if the information about crypto-currencies is stored on plastic media for some time. We can at least hope that new envelopes will be delivered by mail and not received at bank branches.

How many years will it take before the paper money is completely out of circulation? This is not going to happen before the Internet becomes available anywhere in the world, which, in turn, may happen after the launch of satellite Internet. Both cash and bank cards will most likely disappear at once.

Business as a Whole

What will be the business of the future? What will fundamentally distinguish it from today's business? Universal automation and robotization will be a large-scale event in the next 20-30 years. In addition, virtual space will be developing, so that new types of business will focus on these areas.

People who founded modern large and well-known companies 20-30 or more years ago began to do it when the concept of the Internet was just emerging, when cell phones seemed something unusual, and self-driving vehicles were associated with space technology. Even the founders of today's largest IT companies did not anticipate back then where progress would lead us. Most entrepreneurs just started implementing ideas, and with the advent of new technologies, some of their ideas developed even further, others died out. In many instances, rather large and stable enterprises disappeared from the market within months after the mass launch of new technological solutions.

Very few entrepreneurs are able to quickly understand not only the threats of new technologies, but also the opportunities. People hardly change their beliefs and perceptions, expand the picture; and it can be twice as difficult for a person who has already achieved something and perhaps not so young.

Clearly, any progress spreads among young people much faster - they want to achieve everything as quickly as possible at the lowest cost. The older generation gradually gets used to waiting, suffering, patience and compromise. Whereas when you have a lot of desires and little time, you are ready for experiments, even if they involve an increased risk. And that is why, for example, the overwhelming majority of new IT companies will focus primarily on a younger audience and be created by young entrepreneurs. In 20-30 years, large companies with a management team consisting mostly of young people will remain on the market. We will not try to give the exact age and distinguish between the young and the old. Youth is determined not so much by age as by chemical processes in the body and intellectual flexibility.

The ideas shown in the previous chapter are not unique and can refer to any kind of business and period of time - the progress stands, obviously, for young people. However, we should remember that when young people start doing something, they primarily use the knowledge base, available to them within their lifetime, the attitude, the logical reasoning, and the understanding of the world order. In fact, with all this knowledge in mind they make a decision to act. Therefore, all new enterprises in the next 20-30 years will be based on the absolute confidence of their creators in eternity and universal availability of high-speed Internet, and thus, on the possibility of instant access to any array of information from anywhere in the world.

As it has already been mentioned in the previous chapters, IT will be the key industry in the coming decades. If today IT mainly provides automation services to various enterprises or creates quite universal products, then very soon we will see a completely different picture. IT companies will offer a variety of highly specialized products, new solutions will enter the market, and the perception of possible approaches to meet our needs will also change. What's more, these IT companies help to identify new needs that we were not aware of because of the habits we had over the decades.

We can take financial institutions as an example to demonstrate that a fresh perspective and absence of the necessity to look back enabled using IT to significantly accelerate many procedures that were previously extremely slow. However, that was unlikely to be noticeable in everyday life, as it was a fairly widespread phenomenon. In the telecommunications sector, we also see young mobile operators who somehow manage to capture a significant market share and offer customers a wider range of services at lower prices while they do not have their own infrastructure.

Thus, over the next two or three decades, various IT start-ups will emerge in all sectors of the economy, questioning the efficiency of existing processes and offering new approaches to their modernization and acceleration. "Why so slow?" will be the first issue addressed, followed by "Why then is it necessary at all?" These two simple questions will change all industries, destroying everything slow (and thus unnecessary) and creating everything fast (and thus appropriate). The higher the speed of information exchange, the faster the corresponding business processes, otherwise the information will become obsolete.

In each industry you can easily find many new services that will be demanded as technology develops and consumption patterns change. However, if we look at the gradual transition to virtual reality, we can predict that at some point the market will need to create artificial virtual worlds to address a wide variety of problems in all areas of our lives. In fact, all the time that a modern user spends in the virtual world of communication using smartphones, tablets and personal computers, will be spent in different virtual worlds in the future. Some of them will be completely fictional, having nothing to do with reality, others will be completely realistic, and there will be the combination of these two types.

Based on this, we can expect the emergence of new types of IT companies - a symbiosis of design studios and software development companies (including those from the gaming industry), which will create whole virtual worlds or parts of them, depending on their specialization. As people delve into virtual reality and the market fills with good offers relating virtual worlds, we will spend more and more time there. That means that we will have necessary preconditions for building real financial relationships in such worlds.

From the user's point of view, everything that happens in the virtual world will be quite realistic in terms of using certain results of being in this world and in other worlds or even in real life, because information remains the product of any virtual reality. And information in our understanding is absolutely universal from the perspective of the possibilities of its further application.

Now it is very difficult to imagine everything we are talking about, but in 30 years we will probably change several different virtual worlds within one day, solving problems or spending free time there. So, we expect that the companies mentioned above will be in huge demand.

Financial relations emerging in the virtual reality of the future may lead to the creation of different currencies and full-fledged businesses, which activities relate to the processing of all kinds of information. Developed virtual worlds can be so independent from the rest of the information space that they may turn into real countries with their own regulations and laws. Nowadays, we face similar things in the game world, where players can use real or quasi-money, and companies can promote their brands. But in 30 years everything will be for real - whole worlds with real companies and people inside. Over time, the government with its own laws and taxes will move to these worlds, but so far this is a very distant prospect.

Military

Unfortunately, wars have always accelerated the technical progress. The arms race, even in relatively peaceful times, provides the basis for further rapid technological development in many other sectors of the economy. Certainly, many advanced developments in the military sphere are kept secret for obvious reasons, but we can still make assumptions about the main trends.

For quite a long time, many states have been thinking about eliminating the need for living people to be involved in combat operations by releasing various types of unmanned robots to the battlefields instead. Robotic technology is being developed for all types of military industry. However, since it is not always possible to do without living people, the second task is to minimize the possible threat to their lives and health as a result of participation in military operations. This work focuses on creating protective cyber suits that increase the chances of survival during fire contact.

In 20-30 years, the world's strongest armies will be able to minimize the number of people involved in local armed conflicts. Next-generation unmanned robotic technology will have much greater autonomy and will save people from contact with the enemy.

VR technology will enable the military to be present directly in the fire zone without risking their people's lives, using drones for surveillance and, if necessary, taking over the control and coordination of the entire unmanned group. However, this will probably happen only if individual elements fail after a fire contact.

The equipment will be so powerful that a combat drone soaring in the air over the center of the conflict will suppress all firing points within its visible range without the risk of being destroyed (unless it meets an enemy's superior drone). The conflict will be suppressed not by the scorched earth strategy, but with high precision, classifying all targets and automatically selecting weapons to be destroyed. When using a group of drones, individual units will automatically exchange information on new targets and distribute it within the group.

Returning to the virtual reality, it should be noted that virtual worlds designed specifically for military training purposes will be very useful for practicing various combat situations. By modeling, in the finest detail, the entire neighborhoods or even cities where an attack or defense may occur, including their inhabitants, such worlds will enable the military to try out a variety of strategies and tactics, using the entire weapon arsenal (including nuclear), imitating absolutely realistic consequences of a given action.

In addition, virtual and augmented reality will become indispensable for conducting special operations, which, for some reason, will involve people of flesh and blood. Silent interaction capabilities, in particular, will greatly enhance their effectiveness.

Security

For you and me, as ordinary citizens, safety is, first and foremost, keeping our children safe - wherever they are - without fear of losing them. If our children begin to move freely around the yard, the district, and then the city, it will mean that the security system has reached a fundamentally different level. Are we really so far from achieving this goal? I guess, that in 20 years, in all developed countries, the security issue will be resolved once and for all.

By this time, all cell phones will have had a fingerprint scanner and other types of biometric authentication (at least voice and face recognition). Also, all high-risk devices will be supplied with this technology. All unmanned vehicles in the streets will be equipped with cameras. Thus, cameras and biometric authentication systems will become an integral part of the infrastructure in the coming decades. At the same time, the growing artificial intelligence will be surprisingly fast to combine all available data, not only to find the sources of already committed offences, but also to foresee the very threat of such acts.

It should be clearly understood that even nowadays some intelligence units can prevent offences very quickly (when the infrastructure allows it). But the intelligence services are intended to solve local problems and cannot provide protection of the city population, let alone the country. Moreover, they often act outside the law, and the results of their activities cannot be officially used in the future. Naturally, no one will talk about such capabilities of the intelligence services.

However, with the development of the above-mentioned infrastructure, with the modernization of the legislative base, ordinary law enforcement agencies will provide security in cities quite effectively. But then we need to consider the fact that very soon all devices with cameras that we have will potentially work for a new security system. Surely, new laws will be drafted and adopted, which may enable using certain video recording formats to find violators and prove their guilt. In other words, our cell phones, laptops, tablets and all other devices with cameras will have the legal power to constantly monitor us and record everything that happens. This will be a real shock for many.

Some people have been covering their laptop cameras for a long time and perhaps they should think about covering cell phone cameras. It seems to me that a cell phone witnesses much more interesting moments from its owner's life than a laptop. However, phone manufacturers will not let us cover the cameras so easily - we already see that modern smartphones strongly recommend not to close the cameras for some reasons. By the way, in addition to cameras, many of our devices have microphones, for which a sticker is not an obstacle.

Don't worry too much, though. If we progress along the path of chipping, then camera recording will be trivial compared to the capabilities of biochips embedded in the brain. I don't think that brain will be massively chipped in the next 20-30 years. However, when this happens, the state will be able to read even our thoughts. This is not phantasmagoria. This is much simpler. Most people on the planet constantly use internal articulation when thinking (I would assume that there will be less than a million people in the world, not counting those who can't talk, who have another thinking process). When we read, write or think, we articulate everything in our heads. And, as we remember, this feature will be used to develop new virtual communication technologies.

Unmanned police vehicles will dramatically improve safety in the streets. Communication speed between vehicles far exceeds the information exchange speed between ordinary people. Several cars in the area, moving along optimally calculated routes, combined into a single infrastructure with all cameras, microphones and sensors around, with any smartphone connected to the right place, will make the area completely safe. All policemen will need to do is to have a coffee and enjoy the view from the window. The car will perform everything - in fact, it will be a powerful computer on wheels.

The most valuable thing we have, besides the opportunity to live, is information. Not everyone can realize it, but access to certain information makes us who we are. The ability to access our personal data makes us very vulnerable. Now, when we are completely moving to the information space, it becomes quite obvious. Previously, we could hide some information about ourselves physically, taking advantage of the lack of centralized storage and the imperfection or total absence of information technology. In the future, this will be impossible.

In 20 years any information about movable or immovable property and other assets will be just a record in a certain information system. By changing this record, any person can be deprived of their property. Similarly, by erasing or altering other personal information, it will be possible to change or destroy a person's identity. A person will exist physically, but his personality will no longer exist in the information space. Therefore, physical survival will be a matter of time. After all, all services of the future, as well as money, will be electronic. By today we have almost given up our paper-based passports and other physical identity cards. Our identity in the world of information systems will also be just a record.

On the other hand, various organizations (both state and commercial) are facing the need to store large amounts of crucial information accumulated in the process of doing business or providing services to the public. If information is lost or compromised, the business may be destroyed, bankrupting its owners or paralyzing the activity of a public body.

Obviously, in the coming decades information security will play a crucial role in our society. Very soon, we will see an increase in the number of relevant professionals in government structures, the emergence of new units in the police and intelligence services, a growing demand for information security contractors in the commercial sector and, ultimately, an expansion of information security functions in large companies.

The development of virtual reality will lead to further transformation of information security. It will have to protect virtual worlds from unauthorized access, including attempts to change the worlds themselves or stay in them illegally. Information security specialists will face a new circle of fascinating and unusual tasks. Intruders will seek to steal not only pieces of information, but also entire virtual worlds containing various digital assets.

Space Exploration

Since the first spaceflight in the fifties of the past century, we have not made much progress in space technology, in layman's terms. Satellites are still flying (more modern, of course), and very rarely - people. The astronauts have not been able to get farther than the Moon yet; and that happened very long ago. So far, all space exploration is taking place fundamentally in the Near-Earth space. We have not even reached the borders of the Solar System.

Today, in the vast majority of cases, the space industry continues to use technologies developed more than fifty years ago during the large-scale arms race of that time. Most important space discoveries were made even earlier. During this time, two or three generations of people have grown accustomed to more technology-intensive ways of interacting, and new patterns on consumer behavior has changed the industry. All these factors enabled access to space from a new perspective, perhaps in some sense, to treat it as a consumer. Human mission to Mars and beyond is no longer a product of imagination or science fiction. Space Tourism dream has edged towards reality due to public interest and public welfare.

Whereas before, prominent scientists, the smartest designers and the state were involved in space technologies, now the focus started to shift, and the experience gained over many years of study has fallen into the hands of entrepreneurs. This fact will determine the future of space exploration. The symbiosis of entrepreneurs and talented scientists of our time will help to find quite unexpected applications for seemingly insignificant discoveries. At the same time, the insatiable nature of entrepreneurs will induce scientists to go beyond their assumptions and current ideas, to question the unshakable axioms of the past and, possibly, to discover something fundamentally new that was previously hidden.

I would certainly like to write that in 20-30 years we will have a global breakthrough in space technologies and private flights to the nearest planets will become something commonplace. However, this is not the case. Instead, only spacecraft will fly into space, collecting information that will be used in our virtual worlds and conducting further research. Commercial flights to Earth orbit will become a reality for rich people, but nothing more. It will be funny, if the scientists, who are now seriously talking about the colonization of Mars, suddenly discover that due to the depletion of resources, our civilization emerged from Mars, recognizing the Earth as the best planet for life.

The arms race, which continues today, involves, among other things, missile weapons. Certainly, in 20-30 years, missiles will be much smarter, faster and more maneuverable. New technologies that have to be tested in this field will create more powerful and economical engines. This, in turn, will contribute to the evolution of commercial flights to the near-Earth space.

Development of VR technologies will arouse the interest in space exploration, both from an educational and entertainment point of view. Fitted with powerful video equipment, optics and other electronics, new satellites will enable travelling in space in real time, looking around as far as the eye can see. It will be possible to monitor the flight to Earth's orbit in minute detail. The satellite Internet will provide a full-scale video broadcast. With such technologies, it is quite possible that we will be able to see the nearest planets where our spacecrafts arrive.

Then virtual world creators will commit themselves to work and give us a chance to return to space and other planets as often as we wish. Although we will not reach the real colonization of Mars in the next 20-30 years, we can succeed in virtual worlds.

To conclude the chapter, I would like to note that wireless electricity technology may well catalyze further development of our Galaxy, when near-Earth satellites transmit not only information but also energy. However, who knows whether this will happen in the period under consideration.

Other Industries

In the previous chapters, I have tried to cover all the main aspects of our lives and each economic sector, but nevertheless, much has remained unaddressed.

As for such industries as mining and processing, they will also become fully automated and robotized. There will no longer be anything for humans to do either in the bowels of the earth or at the machine tool. Meanwhile, 3D printing will become the key standard in manufacturing and will evolve rapidly. Eventually, 3D printing will be used to create both the manufacturing equipment and the final products. Using artificial intelligence, we will gradually get something resembling self-reproducing technologies.

Water supply and sanitation will require re-equipping of a whole infrastructure system, and this will take quite a long time. While states are unlikely to meet this challenge in 20-30 years, all settlements of the future will be provided with more modern equipment, assuming automatic maintenance. Old facilities will be gradually upgraded or completely decommissioned. The usual water and sewage pipes may no longer be required, as the new buildings will be built with insulated spaces for water supply and drainage directly embedded in the supporting structures. This will eliminate any risk of leaks and make all assembly and disassembly unnecessary.

In developing countries, waste collection and recycling will finally become a very important issue. Developed countries, however, already have a great deal of experience in this area, so, full automation, robotisation and gradual waste reduction will be the key tendency of the next few decades in this industry. We should keep our Planet clean so that new generations can spread out widely and young people may feel it much more intensely. At some point, a massive campaign will start eliminating the consequences of garbage dumping, but that will require profound change at the government level in developing countries. Twenty years should be enough time for this change.

All professional and scientific activities will be under tremendous pressure from the new technologies. Nowadays many professionals are staying afloat only by maintaining familiar business processes in government and business, and immature legislation on automation and robotics. Mass use of self-learning automated systems is on its way, and this will completely redesign all scientific and professional activities. This will not, however, be a crushing blow to the people involved. Society seems to be adjusting to this change. The last experts and researchers of our time will retire in 20 to 30 years, and current education is already re-engineering its processes so that the number of future scientists or true professionals in the various fields is gradually decreasing. Therefore, the new generations will easily rely on technology, and the states will issue the relevant laws.

Nevertheless, science will, of course, develop at a rapid pace. Only those people will remain in science who really have a talent for it and for whom science is the meaning of life. New age scientists will need to adapt rapidly to developing technology and find ways to apply it to further research. And the states capable of building the required infrastructure that may allow scientists to achieve the highest concentration on the goals will then attract the most promising experts from the rest of the world.

All administration over the next 2-3 decades will also be transferred to automated systems.

We cannot avoid discussing the sublime as well. How will art change? How will the public attitude towards art change? Virtual reality will enable the new generation of artists, sculptors and all fine arts professionals to use new means in order to express themselves. Admirers of these art forms will have access to watching artists in action, connected to properly equipped studios. Workshops will become more common. Creators of virtual worlds will actively cooperate with representatives of the visual arts world to shape the necessary images.

All kinds of performances, as we may remember, will greatly benefit from the advent of VR technologies, as the latter will make them even more spectacular and give the viewers the opportunity to be literally inside the action, viewing it from all angles.

Virtual reality will not provide us with fundamentally new ways to perceive music, but learning music may become easier. As the technology develops, we can create virtual orchestras, ensembles and music bands with all their members staying in their equipped rooms anywhere in the world, while listeners and spectators enjoy their joint work sitting at home on the couch.

The Role of the Government

Most citizens in developing and developed countries still do not have access to quick exchange of information with the government. Will the state of the future follow the lead of the global technological change, or will the new technologies themselves be the result of a certain government impulse? Obviously, the answer to this question depends strongly on the state of development of a specific country. In any case, it will become increasingly difficult for developing countries to maintain at least the current level of lagging behind developed countries.

Not long ago we discussed unmanned transport and the powerful impact its launch would have on all other areas of the economy. But at the same time, we should discuss another issue. If now you go out into any city street in any country and stay there for a while, then in a relatively short period of time you can count quite a few taxis passing by. All these cars do not move by themselves, of course; ordinary living people are behind the wheel. What awaits all these people, when everyone switches to unmanned vehicles? Any country can have dozens, hundreds of thousands, even millions of drivers. What will happen to them when their services are no longer needed? What kind of work will they find, considering that other areas will be actively automated and robotized?

Developed countries with higher standards of living and less manual labor should now find themselves in a more favorable environment, as they had more time to focus on employing people out of work due to various kinds of automation. Some people have already been retrained for new occupations, and many more will have time for retraining in the near future. One way or another, we already realize that manual labor will soon disappear, and we should change and develop to survive and succeed.

The countries that are now leaders in automation and robotization have already begun to realize their superiority and will now be able to take care of their citizens. They will help people find their own applications in developing countries by implementing proven technologies and transmitting new standards for automation and robotization.

Developing countries that have no plans in creating their own new technologies by automating and robotizing all major sectors of the economy risk losing the opportunity to ever cease to be a "developing country." The attempt to use foreign technology will lead to an absolute eternal dependence on other countries.

Those states, which have developed and are still developing due to a considerable amount of energy and natural resources, will face a new reality when resources alone may no longer guarantee economic stability, as new technologies provide energy in sufficient quantities in other ways.

The "manual labor" term today does not have the same meaning as fifty years ago. Today, even an office worker who prepares financial reports can and should be considered a manual worker. So can a taxi driver. All current algorithmically formalized activities will sooner or later be considered manual labor, if performed by a living person.

The longer developing countries postpone the launch of robotization and automation, the more they will lag behind technology and living standards in developed countries. However, to reach mass automation, it is necessary to decide first what to do with notional "taxi drivers".

Very often we judge a country by its capital and make premature conclusions regarding the general standard of living. However, in any country, and especially in developing ones, you can always notice a significant advance in the development of the capital region. Such rapid progress does little to reduce the development gap between this country and other developed countries, because for residents of other regions, any automation or robotization can mean much less than basic amenities. Nevertheless, capitals will be the first to adopt new technologies, which may lead to even greater stratification of living standards even within the same country.

At the same time, we can hardly expect less developed regions to start actively introducing new technologies. Once automation and robotization have started in capitals, developing countries should accelerate the above-mentioned process of reurbanization, as described in the "Real estate and construction" chapter. Conditions should be created in such a way that the residents of capitals and large cities may at some point decide to live in the country, having upgraded all the processes there, having achieved comfortable living in remote regions using advanced technologies. While the regional residents should, in fact, switch from the former manual labor to the current manual labor, replacing those who leave the capitals.

Certainly, the above-mentioned solution may require extreme efforts and measures from developing countries, but if they do nothing, they will most likely have to wait a long time for the next chance to break out of technological, and thus economic slavery.

The introduction of technologies that help to include the residents of remote regions in various economic processes (involving them in remote work, primarily through distance learning) will activate the entire dormant human potential of the country and accelerate reurbanization. Besides, this introduction will attract those people who are less actively ready for the development to cities. What's more, this will help to distribute economically active and educated population and technologies more evenly across the territory and to concentrate economically inactive (or less active) population in urban areas. Naturally, city residents, who are at risk under the implementation of new technologies should be able to choose in advance a group they want to belong to. Distance education or professional retraining will give an opportunity to change people' lives.

What should a state ultimately look like in order to switch from a "developing" to a "developed" country? And what will the developed country be like in 30 years?

Obviously, the automation of all routine operations will help to focus on strategic tasks. Therefore, all services that the state can deliver to its residents will be provided remotely and superfast. Just in case, I'd like to re-emphasize that the concept of document management will become outdated forever. Information systems will gradually replace all kinds of lawyers, economists and other bureaucratic specialists, ensuring processing and comprehensive analysis of various information. The government, in fact, will turn into a large-scale IT-company which works with large data sets and provides its leaders (heads of state) with all the information required for strategic decisions.

Governments will focus on providing their people with a global vision for the next 30-50 years, considering further technological development and geopolitical environment. Developed countries will be led by people with excellent knowledge of both technology and management. These people will be guided only by logic and cold calculation, based on the current needs of the majority, when making decisions affecting the interests of citizens. The government's work will be coordinated by artificial intelligence, preventing emotional decisions.

Let us finally get back to our "taxi drivers". What will happen to those people who, despite all the attempts by the Government to convince them of the need for self-development, choose to stay inactive and cannot to realize themselves either in the real or in the virtual world?

By the time the economically inactive population accumulates in urban areas, technologies will have already gone far away, and almost all the above-mentioned will be implemented and actively used. On the other hand, the development of virtual reality and the gradual migration to the virtual reality will significantly reduce the need for physical goods and services, replacing them with virtual ones. This means that all basic needs of society will be met at a much lower cost. Thus, alternative energy, automation and robotization will be able to provide all people in the state with the minimum necessary to support life, even if most of the society do not work.

At the same time, medicine will gradually lead us to the understanding that chipping of individual organs is an absolutely normal practice if we want to live longer. After that, a turning point will come when certain representatives of the younger generation want to fully merge with virtual reality. Actually, there will be more and more arguments for this (from the limitless possibilities of the brain to the maximum extension of life). At the same time, states will have to minimize costs for economically inactive population in the most natural way, without limiting its rights.

It is easy to guess what will happen after that. Mass brain chipping will be a logical solution to several challenges facing the society of the future at once. After that, the government will have potentially unlimited control over the people who implant chips into the brain. But it will take many years before this process becomes mandatory for everyone. Thus, this point will certainly go beyond the time frame under consideration.

To conclude this chapter, I need to add that moving to the virtual reality will give everyone another chance to start from scratch, as physical assets in virtual worlds mean little. Smart, talented, creative and hardworking people will be able to achieve any results that can be converted into physical assets. What is more important, many people will have the opportunity to devote their lives to science or art, regardless of highly questionable traditions of modern society, which are reduced to consuming as much as possible.

The Virtual Worlds of the Future

So how do we get into virtual worlds? The creation of fictional reality using available information technology no longer seems as challenging as before. Even modern computer games provide excellent representation of the world around us. But how do we really get there, into those fictional worlds, to feel the same way we do in the real world?

Before we answer this question, we need to make one important retreat. These days there is a definite distinction between virtual reality and augmented reality, which implies either total immersion in a certain environment or the superposition of various fictional world objects on reality. At the same time, both of these terms assume, in one way or another, that our consciousness remains in its usual fully aware state and we clearly understand certain limits of the imaginary world, no matter how many signals from various sensors and devices try to convince us to the contrary. Meanwhile, the more realistic the experience we want to get when immersed in virtual reality, the more sophisticated the technology around us should be in order to enable a meaningful interaction with our body's response to what is going on in the fictional world. More simply, when we must run, our feet must actually run if we are conscious.

When we use mobile phones, computers, virtual reality helmets, TV glasses or even specially equipped booths with all the special effects, we are still conscious and able to feel that edge beyond which we can no longer find ourselves. And if we go further along the lines of creating more and more sophisticated devices that interact with each of our senses in an attempt to create the full-fledged reality, their cost will probably make the full immersion technology unaffordable for the vast majority of people.

Thus, we need to think about some alternative. Here again, we are going to deviate a little bit. For over thirty years now, there have been various scientific studies (which are not widely reported for some reason) of a phenomenon called "lucid dreaming". The essence of it is that some people, under certain conditions, are able, during sleep (that is, during the phase of REM sleep, when we see dreams), to become aware of the very fact that it is exactly a dream, without waking up at the same time. Once a sleeping person is aware of a dream, he or she is usually able to control the dream by changing it in a completely random way.

In general, in some sources it is still possible to find the information about various peoples who have long understood this possibility and used it according to their customs and religions. However, you and me, of course, can only be interested in the scientific approach. The first official experiments on the subject began in the late eighties and early nineties due to the devices that could monitor brain activity.

I will not go deep into details but I'd like to prove the possibility of this phenomenon, since there is nothing extraordinary in it and the vast majority of readers, digging in their memory, can flashback, at least, one similar case from their own experience. The main problem for now is that it is somewhat difficult to achieve the moment of awareness itself and then to maintain concentration within a conscious dream for a long period of time. However, with some practice, one can gradually overcome this difficulty.

Some small companies and individual enthusiasts are already trying to develop electronic devices that could increase the possibility of self-consciousness in dreams. And I am absolutely certain that in the very near future, there will be a full-fledged technology that may allow anyone to have only conscious dreams. And this technology will completely change our world.

Why are we deviating in this direction? Because in my opinion, the phenomenon of lucid dreaming is the key to moving into a full virtual reality with no limits. People who have serious experience in achieving the states of awareness during sleep use this opportunity to create their own worlds, in which they can change absolutely everything, including the laws of physics. Sometimes these worlds can be reproduced over and over again, and sometimes they are completely and forever destroyed upon awakening. This also depends on the experience. What's more, the reality of lucid dreaming is no less, and sometimes even more, than the reality of our ordinary lives. I am not going to go deep into details and tell you about those emotional states one can experience in the world of lucid dreaming. However, I'd like to note that in the virtual world, it is apparently possible to realise any desire, whether attainable or unattainable it is in the real world. That is why the above-mentioned technology of dream consciousness can become so widespread and rapid.

Before moving on, let's ask ourselves: who, first of all, right now vitally needs the ability to see lucid dreams? If we had the technology to provide sleep consciousness today, it would change the lives of an enormous number of people who are currently forced into a sedentary lifestyle due to older age or various disabilities. The human brain hardly loses its abilities throughout life, except for specific illnesses or injuries. The overwhelming number of elderly people have quite a working brain until death, and this brain is forced to idle due to the gradual failure of other organs and the consequent inability to fully move and live.

I simplify my reasoning on purpose, so that we can quickly move forward, but I do not write something that cannot be easily checked using open sources of information. So, nowadays, people who lead sedentary lives and are unable, for a variety of reasons, to cause and concentrate on lucid dreaming on their own, might be most interested in experiencing it. Even those people who can induce conscious dreams on a regular basis would appreciate having a simple tool to obtain regular results regardless of their general body condition, including fatigue.

Sleep consciousness technology using external hardware will be only the first step towards complete virtual reality. Among the elderly, there are some people whose brain has partially lost its abilities due to various diseases. Some of them, being in the early stages of progressive diseases, will volunteer to test new technologies that can compensate for the gradual loss of activity of certain parts of the brain. Here we come back to chipping. Biochips embedded or grown in the brains of such people (some of whom may be very wealthy and very progressive, despite their age) will not only compensate for the gradually declining functions of the brain, but also expand them by connecting directly to the virtual world of communication (the Internet, if you like). And this will be the second step into the full-fledged virtual reality. The third and final step would be to merge the two technologies into one.

Naturally, I have described here only one of the possible ways of technological evolution that somehow brings us to the dawn of the new era of virtual reality - it seemed the most logical and understandable to me. Many elderly wealthy people will seek to extend their youth in any way they can. Some of them will only need to have a new heart or liver "printed", while others will face the necessity of brain-based intervention. However, the first volunteer results will start a full-scale transition of the whole society into the new virtual reality.

Let's note that implanting chips into the brain will enable us to choose between different control modes - ordinary reality, augmented reality or complete virtual reality, once the brain is appropriately adapted and trained to work alongside the new technology. This means that perfectly healthy people, once the technology is introduced, will be able to live a normal life as before, but without the need for any mobile phones, TV glasses or other similar devices.

As the number of elderly people in the world increases and many have managed to save money for their old age, this technology will spread rapidly, so there will be no problem. In addition, many people, including very young, will actively seek out the new technology and join the ranks of volunteers. Once that happens, all kinds of virtual worlds will be created (we've already touched upon this topic when discussing business in general), which can now be used in dreams for the effect of total presence, with a tactile perception channel connected, without any limitations or reservations.

Also, these will no longer be dreams in the classic sense (the fast asleep phase doesn't last long enough to be used for long journeys in virtual reality). I believe that due to technology the body will be able to go into a kind of trance or a state of complete relaxation. At the same time, our consciousness will switch completely to a virtual reality mode at that moment, not only at night but also at other, arbitrary times. People will be able to connect the available virtual worlds and spend as much time as required in them.

As for now, it's hard enough to imagine all the possibilities of such worlds, but they will be quite rich. Surely, it will be possible to travel, engage in art, science, sports, and whatever else, as everything will depend only on the capabilities of the virtual world. People entering virtual space by doctor's appointment, of course, will use virtual worlds, to keep their brain and muscles in good condition, to influence certain organs, providing required tension or, on the contrary, relaxation.

Speaking generally about the virtual worlds industry, I believe that entertainment and cognitive topics will be developed first. However, due to literally "absolute" reality in such worlds, the society will gradually reflect on the possible transition to the virtual space for work, training and later, maybe, even for living in it for some periods of time. The more automated and roboticised the world becomes, the more and more people will inhabit virtual worlds. In addition, being in a virtual world is much cheaper, because the body demands less as compared to the real world, especially when it comes to conditioned sleep. In addition, the body can rest all the time, and controlled signals from the brain can provide the necessary contractions to maintain muscle tone. In other words, when training physically in the virtual world, you may well provide the required load for the resting body so that the muscles do not atrophy from long periods of rest.

When it comes to jobs, virtual reality will offer people from all over the world the chance to work in the same office from home, or, like a chess player playing multiple games at once, to be in multiple places at once.

Those companies that manage to occupy the niche of creating and developing virtual spaces will continue rapidly gaining momentum. Huge virtual worlds will begin to appear, with their own laws and no visible limits in their development. People would no longer care about the density of population on our planet - there would now be enough space for everyone.

And then at some point people will realize that the elderly, who seemed to have been forgotten and rejected by everyone from the real world, will suddenly come back to our world full of energy and new ideas. After all, they will now have access to the same virtual worlds as all other people - virtual reality technology will enable them, among other things, to create any image they want and to regain their youthfulness. And that means they will be with us until the very end - cheerful, motivated and full of energy. And that awareness will definitively change the understanding of the potential applications of the new virtual reality technology.

It is easy to assume that those elderly people, who due to health issues can no longer move a lot in real life, will be unwilling to leave virtual reality after they have perfectly mastered the full range of virtual world tools. But we won't blame them for that, especially if their lives turn out to be happy and fulfilled. Instead, we will design special capsules that can maintain all body functions for as long as possible and see what will happen.

Chipping

We have already addressed chipping several times, but now let's try to tell the whole story here. What will we end up with and how will it be used in everyday life?

We have already mentioned that pets are being chipped in many countries these days, and far from causing any kind of social rejection, on the contrary, it is being legislated. Generally, pets do not live as long as humans, and the organ transplantation technology differs a little bit when compared to humans. And since the technology for printing artificial organs is already being actively developed, we can assume that pets may well become the first volunteers to have certain organs replaced with new, artificially synthesised ones, in cases there are certain medical indications.

These organs will contain next-generation biochips inside enabling them to monitor their performance, as well as other processes occurring in the animal's body. It is easy to guess that it would be incredibly convenient for artificial organs manufacturer to have constant access to the data from biochips, eliminating any need for physical contact with the animal. This, in turn, implies that biochips should gain direct access to the virtual world of communication.

Alongside this process, there has been rapid and constant development in the field of bionic prostheses for humans. Very soon, such prostheses will be equipped with various electronics that may not only provide communication with the human nervous system, making the prosthesis work identically to a living limb, but also provide additional functions. Depending on bionic prosthesis type, its modification and owner's personal preferences, 3D-printing will enable any functionality, turning a prosthesis into a full-fledged functional computer, with an option to upgrade its individual components or, certainly, to replace the entire prosthesis.

In the next stage, we can expect both the fields of bionic prosthesis design and artificial organ growth to fully merge into a single bioprosthetic industry, where a bioprosthesis can substitute any organ, either external or internal. Successful experience of organ replacement in pets as well as the first human volunteers, who have no choice but death or transplantation, will do the trick. Meanwhile, the new artificial organs will become more advanced than any given organs by nature, and will be equipped with biochips to monitor remotely our general state of health. Subsequently, replacing organs with artificial ones will spread widely, certainly affecting older and less healthy people first. At this stage, the biochip inside will become commonplace, but still be associated primarily with medical necessity.

As we see it, biochips will be the result of an immense effort of technological thought, various innovations, research and discovery. Ultimately, as discussed above, a prototype biochip will be produced to be used in the brain both as a prosthetic and communication device at once. I believe that biochips providing us with direct contact between the brain and the virtual world of communication will be available much sooner than in 20 years' time, although some sort of medical reason may be required for their mass use in society. Otherwise, if there is no medical reason, governments may simply not permit human implantation, although, of course, such experiments may be carried out illegally. Therefore, we will wait for the biochips capable of overcoming various pathological age-related changes in the brain to come, in order to test the system's capabilities of establishing direct communication with the virtual world. Even so, I believe we will get our first results within, at the latest, 20 years. Nevertheless, as is mentioned above, mass chipping is likely to go beyond the range under consideration.

As we may remember, biochips implanted in the brain should function as adapters, linking different types of electromagnetic waves and training the brain to communicate with the outside information space. When this happens and the first people have direct access to the Internet, all people will start switching to the new standard of communication. Apart from those people with medical reasons for brain bio-chip implantation, young people from the IT sector will be the first to become interested in increasing the speed of information exchange with the world around them. For young people chipping will be a logical extension of ongoing automation processes in the world. In addition, young people are more at ease with their bodies and have a different sense of personal freedom.

Now, let us briefly list the main opportunities for humanity as a whole which emerge after the transition to direct access to the virtual world of communications. Certainly, the lifetime and especially its active period will increase. Moreover, the society will experience an era of absolute security. In terms of convenience, all devices will be controlled remotely by thought, and this distance will not be limited in any way, except for possible security limitations in the networks the devices are connected to (we will only be able to control those devices that allow us to do it). In general, everything related to electronics will be controlled by thought. Besides, the human brain will gradually develop the ability to manage all the information available in the virtual communication world, just as it currently manages the data stored directly in the head. In other words, we will have instant access to answers to any questions, and we will even be able to use artificial intelligence under certain conditions. The current knowledge test system will have to give the highest grade to all students. That is why, among other things, the education system will be greatly transformed, as already mentioned in the relevant chapter.

We can easily guess that biochips will bring medicine to another level. Either way, special software will gradually train the brain so that all medically important information can be collected through self-analysis and gathering the required data from all organs and systems of our body. I don't think blood analysis will be relevant once our chipped brain learns to determine its own composition. However, such progress will not be visible straight away. At first, automated medical systems, interacting with the brain via biochips, will establish baselines and identify potential risks to be followed by still more familiar treatment methods.

Human competition will no longer exist on the same level. It will enable to identify who is better able to use human brain in managing the shared information space, or who wins in personal decision-making, research and so on. It is difficult to explain in words how we will respond to the continued presence of access to all the world's information. And it is even harder to describe which changes in our brain will follow when forced to adapt to processing such volumes. I'd like to use an analogy to explain my train of thought. The Internet will become some sort of a new subconscious mind for most people, while very few will handle the single information space consciously, feeling its full potential and using it as intended, for progress and creation.

As we may remember, brain biochips will first enable elderly people, and then everyone else, to connect to virtual worlds of all kinds. The biochip will train the brain to enter into a trance state for subsequent transition into various fictional, artificially created information spaces, where one can do anything. These virtual worlds can link any people in various parts of the world with full-fledged tangible connections and provide them with the opportunity to communicate, implement projects together, travel, study and simply live. Once in these virtual worlds, people will be able to develop them, build new ones, and merge with one another. Perhaps one of these virtual worlds will later become an exact replica of our modern world.

As for states, they will potentially have full access to all people's personal and private information after mass chipping. In addition, states may have enough power to change the behavioural patterns of the chipped people and influence fertility and mortality rates. That does not sound very reassuring, but we should understand that people working in national governments will be subject to the rules and will not be able to interfere in other people's private lives simply at will. The system will operate according to the rules of artificial intelligence. Of course, the transition to full chipping will take several more decades, during which technology may take complete control over the human factor, to say the least.

As for various concerns about chipping and its dangers to the public, I'd like to state the following. Obviously, we should be very careful in predicting the consequences of chipping technology for the humanity, especially with regard to the unauthorized use of some of its functions. However, we should also keep in mind the current public enemies represented by drugs, tobacco, alcohol and quite a few others. Who knows: perhaps, chipping will help us to overcome them?

As you may realize, I do not question the possibility of a direct access to the virtual world of communications. Let's think about it: can this technology simply not be used, and people give up using it? Can we imagine an alternative to direct access to the virtual world of communications in 50 years' time? Will young researchers or terminally ill people be able to resist the temptation of fundamentally different ways of interacting with information, as well as unlimited opportunities for self-realization in virtual worlds? Will states or society deny them these choices? And will all other people then be free to live their lives safely aware of other people living more quality lives with no boundaries to self-discovery and self-development?

New Time and Major Issues

This chapter is not for everyone. Some reflections may seem completely unbelievable or even insane. Nevertheless, the rapid development of technology is forcing us to look at the world of information in many new ways, realizing that we ourselves and the space around us are the information.

So we can't help but ask ourselves: will the next generation finally be able to get answers to the eternal questions facing humanity for a long time? Will an understanding of the meaning of existence change after we move to a virtual world, or will the need to comprehend this meaning disappear altogether? Will the way to live forever be invented and what is immortality? Will there be a scientific answer to the question of what comes after death if it cannot be overcome?

We will, of course, try to answer these questions, but as always, the new times will bring other questions to people - maybe not as global, but no less urgent. What and how will they believe? How will religion communicate with people: will it follow them into a virtual world or will it try to pull them out of it? What thing will be the most precious for the individual and for the humanity in a generation? How will the young people of the new age who can't or won't move their consciousness into virtual worlds and want to stay in the real world live?

Let's start with faith and religion. Once people start actively using virtual worlds, no one else can stop them from believing in anything they want, building their own worlds with their own rules, their own religions and beliefs. Traditional religion, for sure, will follow them, but for obvious reasons, it will lag significantly behind in understanding of the ongoing changes. Besides, it even may try to slow them down, as ever before in history. But somehow the only way to fight for one's place in the changing world will be to follow the general trends. So, at some point, religion will follow us into the virtual reality. And by the way, even now we can find the first attempts of religion to penetrate the virtual world of communications in the electronic shops of mobile applications.

Let's proceed to the next question. What will be the most valuable thing in 30 years' time? People who have constant access to virtual reality will have complete self-realisation in it, and it won't take much to maintain a physical body. With a small amount of personal space, a source of energy, water, and access to modern technology, it will not be difficult to stay in a good shape. The new middle class of the coming decades will seek to satisfy the need for its own territory, water and energy. These will be the most valuable assets. These will be all that will remain real in our lives. The rest will be generated in the virtual reality.

Will gold still remain a so-called protective asset, for example? Maybe it will be for a while, by inertia, but the new generation will not be able to understand the yellow metal's protective properties. Will it be possible to exchange it for water at a favourable rate? Some will keep the gold as an heirloom for a long time, but it will be mainly used in industry - the age of the gold standard will finally come to its logical conclusion.

Self-reproducing technology will become the most valuable resource for humanity as a whole. Even drinking water will not be so important when desalination technologies reach a new level. Every country will begin to think about how, after any global war or disaster, to quickly restart all processes and provide the survivors with everything they need. Special protected bunkers will now store not only food for the first time, but also technological laboratories, a kind of mother 3D printers equipped with required consumables, allowing people to reproduce from scratch everything that will be needed. All critical information will also be stored there. Losing access to information will set us back several decades, given the ongoing irreversible processes in the education system.

Turning to immortality, I should say that never before mankind came so close to the solution of this problem. In modern interpretation the immortality of any object is the ability to preserve for an infinitely long time a complete array of the information that the given object is identified with. In this sense new technologies will enable us to create in time some variety of virtual worlds, each of them representing a full copy of all the information required for self-consciousness with all inner history. As a matter of fact, you and I are now each connected to our own virtual world of thoughts, feelings, memories and sensations. And one day technology will permit these worlds to be copied to preserve our essence over time.

As for what happens after death, we are obviously also dependent on previous considerations. When technology allows for the preservation of a person's essence after it is no longer possible to maintain life in his physical body, we will realise that death as such never existed. Instead, information was transferred into another state, which we could not control or understand in any way before. Who knows: may be such rethinking will enable us in some decades to find the traces of our ancestors in the open spaces of information and to revive them.

How will the people of the new generation, who want to stay outside the virtual reality, live? The answer to this question will be very short. They will live approximately the way some tribes in distant parts of our planet live now. This is the only way to keep newborn children from being tempted or having to come into contact with the virtual reality.

We still have to discuss the meaning of existence. Nowadays people have so many different opinions about it, from very concrete to very vague, from absolutely understandable on the physical level to totally unperceivable. But the challenge is to formulate a clear answer to this question, given the realities of what society will face in 30 years, when technology will have completely changed our understanding of information. And the answer is very simple and obvious. Life is about constantly adapting to the changes that occur all the time around us. If we adapt fast enough, we can enjoy it, but if we adapt too slowly, it can only be painful.

Conclusion

Well, that brings us to the end of our short walk through the technological perspectives of the next two or three decades. Perhaps, it was too quick to see the details, but we will soon see all these changes implemented in practice in our real lives.

What conclusions can we draw from the above? The future of all economic sectors and spheres of our life already fully depends on the development of information technology. This means that every governmental and commercial organization needs to focus on establishing completely autonomous IT units, which are the source of all the future development. Any external appeal would risk losing competitive advantage, information or even the loss of the entire organisation. When planning the launch of new business projects, it is essential to consider the maximum perspective of technology development and to build the infrastructure in advance for the subsequent transition to it.

Furthermore, given the development of 3D printing as one of the most important manifestations of technological progress, developing countries with enough natural resources can take advantage of a unique opportunity. In fact, today the technology is still in its infancy and has great potential, and the world still lacks the necessary expertise. A gradual and persistent development of this technology, if only from scratch, could make it at least temporarily independent from external economic factors.

Perhaps, I have forgotten to mention just one important thing in the book. When we start implanting biochips in the brain, humanity will be able to communicate telepathically. And if we used to talk about telepathy as something unscientific or supernatural, now this phenomenon can unfold before us in a completely new, understandable and logical way. Reflecting upon it, it seems that in the near future most of such mysterious or, as stated sometimes, esoteric notions about human life may receive a certain technological interpretation, losing all its mystic charm.

Like telepathy, humanity will have access to another technology so often described in fantasy novels. Once we realize that everything around us is information, the questions of instant movement of any physical object in space will no longer seem unsolvable.

Today, however, we regret to note that the entire global economy has been hit hard during the crisis, partially caused by the public reaction to the coronavirus pandemic. Thus, many countries have lost precious resources that could have been invested in the extensive automation and robotisation of all areas of our lives. At the same time, the progress rate has been slower in recent years and many workplaces have already been lost anyway.

You should never wait for a crisis to accelerate progress. And we should not worry losing jobs once the era of new technologies arrives. It is far worse to lose human lives than jobs. And people will manage to keep themselves busy in the new world.

The time we all spent in enclosed spaces may well be remembered as the very last instance of a massive increase in the number of people on the planet, prior to the gradual and inevitable decline in the birth rate. And the children born in 2021 will be the real heirs to the forthcoming changes.

No one will stop them from discovering themselves, exploring new technologies, making discoveries, building companies, changing our world more and more from the time they are very young. No one will impose their point of view or force them down the wrong path.